**Springer Theses**

Recognizing Outstanding Ph.D. Research

## Aims and Scope

The series "Springer Theses" brings together a selection of the very best Ph.D. theses from around the world and across the physical sciences. Nominated and endorsed by two recognized specialists, each published volume has been selected for its scientific excellence and the high impact of its contents for the pertinent field of research. For greater accessibility to non-specialists, the published versions include an extended introduction, as well as a foreword by the student's supervisor explaining the special relevance of the work for the field. As a whole, the series will provide a valuable resource both for newcomers to the research fields described, and for other scientists seeking detailed background information on special questions. Finally, it provides an accredited documentation of the valuable contributions made by today's younger generation of scientists.

## Theses are accepted into the series by invited nomination only and must fulfill all of the following criteria

- They must be written in good English.
- The topic should fall within the confines of Chemistry, Physics, Earth Sciences, Engineering and related interdisciplinary fields such as Materials, Nanoscience, Chemical Engineering, Complex Systems and Biophysics.
- The work reported in the thesis must represent a significant scientific advance.
- If the thesis includes previously published material, permission to reproduce this must be gained from the respective copyright holder.
- They must have been examined and passed during the 12 months prior to nomination.
- Each thesis should include a foreword by the supervisor outlining the significance of its content.
- The theses should have a clearly defined structure including an introduction accessible to scientists not expert in that particular field.

Indexed by zbMATH.

More information about this series at http://www.springer.com/series/8790

Zhen Shen

# Experimental Research of Cavity Optomechanics

Doctoral Thesis accepted by
University of Science and Technology of China,
Hefei, China

Springer

*Author*
Dr. Zhen Shen
University of Science and Technology
of China
Hefei, China

*Supervisor*
Prof. Chun-Hua Dong
University of Science and Technology
of China
Hefei, China

ISSN 2190-5053 ISSN 2190-5061 (electronic)
Springer Theses
ISBN 978-981-33-4457-0 ISBN 978-981-33-4458-7 (eBook)
https://doi.org/10.1007/978-981-33-4458-7



This Springer imprint is published by the registered company Springer Nature Singapore Pte Ltd.
The registered company address is: 152 Beach Road, #21-01/04 Gateway East, Singapore 189721, Singapore

# Supervisor's Foreword

The optomechanical interactions take place via either the radiation-pressure force induced by the optical fields or processes such as Brillouin scattering. These interactions can lead to quantum state transfer between relevant optical and mechanical systems and can also generate quantum entanglement between these systems, which is hard to get that regime. This thesis reports the optomechanical interactions in a high Q microsphere to achieve non-magnetic non-reciprocity and develop all-optically controlled non-reciprocal multifunctional photonic devices. These unique features establish a new avenue towards integrated all-optical switching with low-power consumption, optical isolators, circulators and directional amplifier. Especially, by coupling two degenerate optical WGMs in a such microresonator to one of its mechanical breathing modes, mode conversion has been demonstrated via the mechanically dark super optical mode in a classical regime. The signatures of the dark mode excitation are no OMIT for the dark mode since this mode is decoupled from the mechanical oscillator. These proof-of-principle experiments demonstrate the special role of the dark mode in the optical wavelength conversion process, opening the door to mechanically mediated optical state transfer in a quantum regime without cooling the mechanical system to its motional ground state. In addition, the single-bit quantum memory based on circulating acoustic phonons has been proposed. All achievements described here are very helpful to study quantum information processing and quantum devices with optomechanical oscillators.

Hefei, China
August 2020

Prof. Chun-Hua Dong

# Preface

Radiation pressure is the pressure exerted upon any surface exposed to electromagnetic radiation. The radiation-pressure force can be used for cooling atomic motion. Laser cooling was experimentally realized in the 1980s and has since become an extraordinarily important technique. Many applications have been enabled by laser cooling, including optical atomic clocks, precision measurements of the gravitational field, and systematic studies of quantum many-body physics in trapped clouds of atoms. Since the advent of both optical microcavities and advanced nanofabrication techniques, the optomechanical interaction could be studied in mesoscopic mechanical resonator. In 2005, Kippenberg et al. discovered that optical microtoroid resonators with their high optical finesse at the same time contain mechanical modes and thus are able to display optomechanical effects, in particular, radiation-pressure-induced self-oscillations. On one hand, this is due to the high intensity local field in the microcavity, which can enhance the optomechanical interaction. On the other hand, the quality and elasticity of the microstructures are small and prone to deformation. During the last decade, cavity optomechanics has advanced rapidly and optomechanical coupling has been reported in numerous novel systems. Furthermore, a number of remarkable optomechanical phenomena have been demonstrated in experiments, such as amplifying and cooling of phonon modes, optomechanically induced transparency, strong coupling between mechanical oscillator and optical cavity, optical wavelength conversion, conversion between optical light and microwave, optomechanical light storage and optomechanically induced non-reciprocity. In future, both in the field of precision sensing, classical and quantum information processing, cavity optomechanics will play an increasingly important role.

Here, we study the interaction between the optical field and the mechanical oscillator based on the whispering-gallery modes microresonator. The main contents include:

1. Whispering-gallery modes microcavity and optomechanical interaction. Whispering-gallery modes microcavity has independent optical and mechanical properties, which will be introduced first as well as the fabrication method of microresonator and tapered fiber. Then based on the experiment of

optomechanically induced non-reciprocal transparency and light storage, we describe the optomechanical interaction in microcavity and related measurement techniques, which are crucial for the subsequent experiments.

2. Optomechanically induced non-reciprocity. Non-reciprocal devices, such as circulators and isolators, which allow light to pass in a specified direction but block light in the undesired direction, are indispensable components in classical and quantum information processing in photonic integrated circuits. However, the integration of the conventional optical non-reciprocal devices, which employ the magneto-optical Faraday effect, is technically challenging because of material incompatibilities and the large footprint of such device designs. In recent years, there has been growing interest in the realization of non-reciprocal photonic devices without magnetic materials. Here, we experimentally demonstrated non-magnetic non-reciprocity using optomechanical interactions in a whispering-gallery microresonator, as proposed by Hafezi and Rabl. In this approach, the traveling-wave nature of the microresonator produces a rise in the degenerate clockwise (CW) and counter-clockwise (CCW) traveling-wave optical modes. For the optomechanical interaction, the CW and CCW modes are independently coupled with the mechanical mode. Only when the optical driving and signal fields are coupled to the same optical mode, the coherent conversion between signal photon and phonon is enabled. As a result, the directional driving field breaks the time-reversal symmetry and leads to non-reciprocal transmittance for the signal light. In a fiber coupled silica microsphere system, we observed optomechanically induced non-reciprocal transparency (OMIT) and amplification (OMIA), and demonstrated a non-reciprocal phase shift of up to 40 degrees.
3. Brillouin-scattering-induced transparency and non-reciprocal light storage. In the previous experiment, we focus on the radiation-pressure-driven mechanical oscillator. Here, we study the optomechanical interaction between the traveling optical and mechanical whispering-gallery modes. We demonstrate the Brillouin-scattering-induced transparency in a high-quality whispering-gallery-mode optical microresonator. The triply resonant Stimulated Brillouin scattering process underlying the Brillouin-scattering-induced transparency greatly enhances the light-acoustic interaction, enabling the storage of light as a coherent, circulating acoustic wave with a lifetime up to $10\mu s$. Furthermore, because of the phase-matching requirement, a circulating acoustic wave can only couple to light with a given propagation direction, leading to non-reciprocal light storage and retrieval. These unique features establish a new avenue toward integrated all-optical switching with low-power consumption, optical isolators and circulators.
4. Packaged optomechanical microsphere. For the quantum optomechanics research, the mechanical oscillator needs to be cooled to ground state in a cryostat or vacuum chamber, which can also improve the $Q$ factors of the mechanical modes. However, stable 3D optical stages are required to hold the microresonator-taper coupling system to keep and adjust the air gap between them and, thus, give rise to experiment difficulties in a relatively small vacuum

chamber or cryostat. To address this issue, an efficient method to mount a coupled silica microsphere and tapered fiber system is proposed and demonstrated experimentally. For the purpose of optomechanical studies, high-quality-factor optical and mechanical modes are maintained after the mounting process. For the mounted microsphere, the coupling system is more stable and compact and, thus, is beneficial for future studies and applications based on optomechanical interactions. Especially, the packaged optomechanical system, which is tested in a vacuum chamber, paves the way toward quantum optomechanics research in cryostat.

5. Phase-sensitive imaging of vibrational mode. The characterization of vibrational modes of electro/optomechanical devices, including identification of the vibration frequencies, profiles, mode losses and so on, is important for validating their design, operation and performance. Scanning optical vibrometer enables high-resolution, non-invasive mapping of the vibration pattern of acoustic devices. It also offers complete quadrature information rather than pure intensity response. We demonstrate a high frequency phase-sensitive heterodyne vibrometer, operating up to 10 GHz. Using this heterodyne vibrometer, the amplitude and phase fields of the fundamental thickness mode, the radial fundamental and the second-order modes of an AlN optomechanical microdisk resonator are mapped with a displacement sensitivity of around $0.36\,\text{pm}/\sqrt{\text{Hz}}$. The simultaneous amplitude and phase measurement allows precise mode identification and characterization. The recorded modal frequencies and profiles are consistent with numerical simulations. This vibrometer will be of great significance for the development of high frequency mechanical devices.

Hefei, China
August 2020

Zhen Shen

# Acknowledgements

I would like to express my gratitude to all those who helped me during the writing of this thesis.

Firstly, I would like to thank Prof. Chunhua Dong, my thesis advisor, for his patience, support and guidance throughout my doctoral research period. His fascination with the frontier of science and deep understanding of physics are most valuable lessons. It has been a great pleasure to be his first student. I would like to thank Prof. Guangcan Guo, the director of the laboratory. Under his leadership, our lab has the advanced experimental platform and free academic atmosphere. I give my sincere gratitude to Prof. Fangwen Sun. I had this opportunity to study here because of his appreciation and recommendation.

I would like to thank Prof. Changling Zou and Dr. Yanlei Zhang for their enthusiastic guidance and support, especially in terms of the understanding of physics, which help my experiments to be carried out smoothly. I want to thank other Professors in our lab: Prof. Xifeng Ren, Prof. Shuang Wang, Prof. Jinming Cui, Prof. Xiangdong Chen, Prof. Deyong He, Prof. Chuanfeng Li for their support on my doctoral research.

I give my sincere thanks to all my colleagues in the lab who gave me generous help, Yuan Chen, Zhonghao Zhou, Shuai Wan, Chengzhe Chai, Xinxin Hu, Rui Niu, Leiming Zhou, Yongjin Cai, Le Yu, Congcong Li, Ao Shen, Xiao Xiong, Aiping Liu, Shen Li, Ming Li, Lantian Feng, Yang Dong, Jian Wang.... I wish best of luck to all of you.

I also express my thanks to lab administrative staff, especially Huai Ye, Kaimin Duan, Tao Wang, Li Wu, Hongmei Li.

Because of the recommendation of Prof. Changling Zou, I had the opportunity to visit Prof. Hong X. Tang's group and study at Yale University. I would like to thank Prof. Hong X. Tang for his guidance and support. I also thank other members of the group, especially Wei Fu. I also visited Peking University shortly. I would like to thank Prof. Yunfeng Xiao. I also express my gratitude to other members of the group, especially Dr. Beibei Li, Dr. Xuefeng Jiang, Li Wang.

I thank my family for their supports, especially my wife, who are extremely important to me. Your expectations are also a great motivation for my study and work.

# Contents

# Chapter 1
# Introduction

**Abstract** Radiation pressure is the pressure exerted upon any surface exposed to electromagnetic radiation. The radiation-pressure force can be used for cooling atomic motion. Laser cooling was experimentally realized in the 1980s and has since become an extraordinarily important technique. Many applications have been enabled by laser cooling, including optical atomic clocks, precision measurements of the gravitational field, and systematic studies of quantum many-body physics in trapped clouds of atoms. Since the advent of both optical microcavities and advanced nanofabrication techniques, the optomechanical interaction could be studied in mesoscopic mechannical resonator. In 2005, Kippenberg et al. discovered that optical microtoroid resonators with their high optical finesse at the same time contain mechanical modes and thus are able to display optomechanical effects, in particular, radiation-pressure-induced self-oscillations. During the last decade, cavity optomechanics has advanced rapidly and optomechanical coupling has been reported in numerous novel systems. Furthermore, a number of remarkable optomechanical phenomena have been demonstrated in experiments, such as amplifying and cooling of phonon modes, optomechanically induced transparency, strong coupling between mechanical oscillator and optical cavity, optical wavelength conversion, conversion between optical light and microwave, optomechanical light storage and optomechanically induced non-reciprocity. In future, both in the field of precision sensing, classical and quantum information processing, cavity optomechanics will play an increasingly important role.

When light strikes the surface of an object, it transmits momentum to the object and generates a force, also called radiation pressure. As early as 1873, Maxwell predicted the optomechanical effects of electromagnetic fields when he established the electromagnetic field equation [1]. However, in the process of experimentally demonstrating radiation pressure, other effects, such as thermal effects, need to be carefully excluded, so the experiment is extremely difficult [2, 3]. Until the beginning of the 20th century, the existence of radiation pressure was experimentally demonstrated by Lebedew, Nichols, and Hull [2, 3].

In the 1970s, Hansch et al. proposed using radiation pressure to cool the random motion of atoms, that is, "laser cooling" [4, 5]. During the same period, Ashkin who



Z. Shen, *Experimental Research of Cavity Optomechanics*, Springer Theses,
https://doi.org/10.1007/978-981-33-4458-7_1

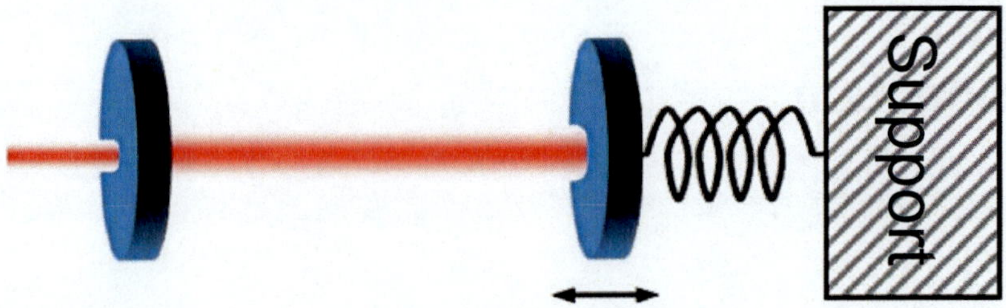


**Fig. 1.1** Schematic diagram of a typical cavity optomechanical system

won the 2018 Nobel Prize in Physics for the optical tweezers and their application to biological systems, demonstrated that light can trap and manipulate neutral particles [6]. In the 1980s, laser cooling was realized and it has become an extremely important technique since then [7, 8]. Steven Chu and coworks were awarded the 1997 Nobel Prize in Physics for their independent pioneering research in cooling and trapping atoms using laser light. Optical force is widely used in atomic clocks, precision measurement of gravitational fields, and quantum many-body physics in trapped clouds of atoms [9, 10]. In particular, atoms or ions that are effectively cooled by the laser to the ground state can generate exotic motional states, such as Fock or Schrodinger cat states [11–13].

The use of radiation pressure to cool macroscopic objects was first proposed by Braginsky in the context of interferometers. Braginsky considered that one of the cavity mirrors was suspended as a mechanical oscillator, as shown in Fig. 1.1. Due to the finite lifetime of the cavity mode, the retarded effect of the optical force can change the dynamics of the mechanical degree of freedom, which could give rise to the damping or amplification of mechanical motion. Initially, he demonstrated this dynamical backaction in a microwave cavity [14, 15]. In the optical domain, the bistable phenomenon of radiation pressure was first demonstrated in a macroscopic optical resonator [16].

In addition to phenomena that can be explained in classical terms, Braginsky also proposed the quantum fluctuations of radiation pressure which impose a sensitivity limit in displacement measurements of the mirror [17, 18]. In this regard, Caves and Jaekel also provided a detailed analysis of the quantum noise in the interferometer [19, 20]. These work established the standard quantum limit of continuous displacement measurement, which is of great significance to the gravitational wave detectors. In 2017, Weiss, Barish and Throne were awarded the Nobel Prize in Physics for their decisive contributions to the LIGO detectors and the observations of gravitational waves.

During the 1990s, theorists began to explore the various quantum effects of cavity optomechanical systems. For example, quantum nondemolition (QND) detection of the light intensity [21] and the generation of squeezed light [22] were exploited using the optomechanical interaction as an effective Kerr nonlinearity. When in the extremely strong optomechanical coupling region, the nonclassical and entangled states of the motional states and optical degrees of freedom could also be achieved

[23]. During the same period, the technology of cooling atomic motion by cavity-assisted laser-cooling was also developing [24].

For the experiments of cavity optomechanics, optical feedback cooling by dynamical backaction of radiation pressure was first demonstrated in 1999 [25]. At the same time, the size of the optomechanical cavity was reduced continuously, such as measuring the thermal motion of a millimeter-order resonant cavity [26]. However, it was challenging to obtain a higher quality factor in a smaller optical Fabry–Perot cavity. Nevertheless, it was possible to observe the dynamical backaction of optomechanical effects at micro scale where the forces originated from photothermal effect, including optical spring effect [27], feedback damping [28], self-induced oscillations [29], and cavity feedback cooling [30].

Because the unwanted heat due to photothermal interactions which tend to mask quantum signatures, the non-dissipative radiation pressure force has more potential for the future quantum applications of optomechanics. Both the advent of optical microcavities and the tremendous progress in micro- and nanofabrication technologies allow the cavity optomechanics to be studied at micro scale and enter the quantum regime. In 2005, Kippenberg et al. first observed the mechanical vibration mode and optomechanical effects in a microtoroid cavity, and realized self-oscillations induced by radiation pressure [31–33]. Figure 1.2 shows the three mechanical vibration eigenmodes of a microtoroid cavity and the resonance frequencies as a function of cavity radius. It can be seen that the mechanical vibrations of the microcavity modulate the optical cavity length, and the power spectrum of transmitted light field generates a series of sidebands. Since then, the field of cavity optomechanics has advanced

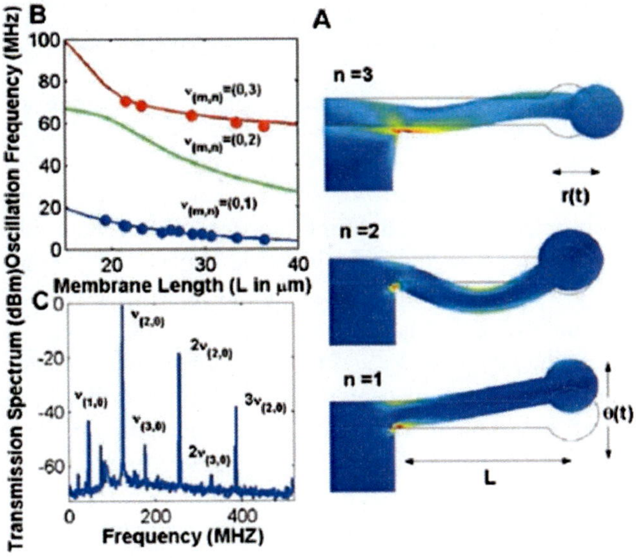


**Fig. 1.2** The mechanical modes and optomechanical interactions in a microtoroid cavity. Reprinted with permission from Ref. [32]. Copyright 1998 by American Physical Society

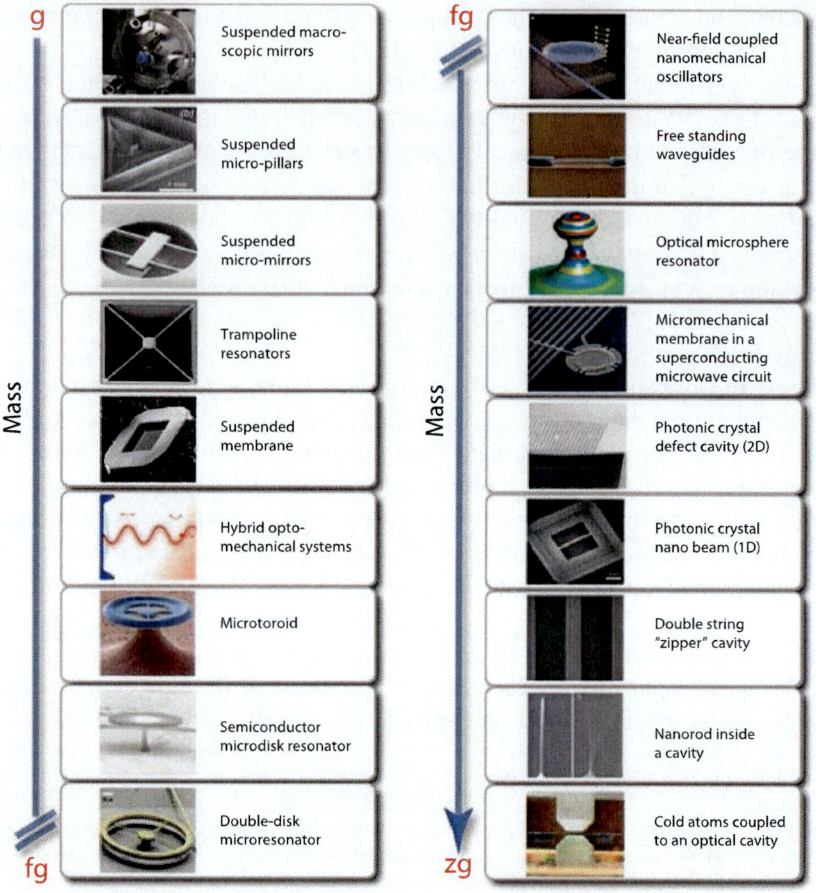


**Fig. 1.3** Optomechanical devices of different mass. Reprinted with permission from Ref. [75]. Copyright 1998 by American Physical Society

rapidly and the optomechanical coupling effects have been reported in various systems, including membranes and nanorods inside Fabry–Perot resonators [34, 35], whispering gallery mode microdisks [36, 37] and microspheres [38–40], photonic crystals [41, 42], and evanescently coupled nanobeam [43], as shown in Fig. 1.3.

In 2006, optomechanical cooling was realized in both the microtoroid resonators and the suspended micromirrors systems [44–46]. In microtoroid cavity, Schliesser et al. exploited optomechanical cooling to cool a mechanical mode with frequency of 58 MHz from room temperature to 11 K [46]. In 2008, they first demonstrated the resolved-sideband cooling in mesoscale microtoroid resonators [47]. The linewidth of the optical whispering-gallery mode used in the experiment was only one-twentieth of the mechanical mode frequency, and the cooling rate exceeded 1.5 MHz, which was three orders of magnitude greater than the inherent dissipation rate of mechanical mode. In 2009, they presented measurements on optomechanical systems exhibiting radiofrequency mechanical modes of 62–122 MHz, cooled to occupancy of around 63 quanta using a combination of cryogenic precooling and resolved-sideband laser cooling. The cooling laser simultaneously acted as a highly ideal displacement transducer, demonstrating the closest approach to the Heisenberg uncertainty product for continuous position measurements [48]. Around the same year, Park et al. reported mechanical oscillations of a deformed silica microsphere are coupled to optical whispering-gallery modes that can be excited through free-space evanescent coupling. Combing precooling the system to 1.4 K and resolved-sideband laser cooling, a final average phonon occupation as low as 37 quanta was achieved [39]. In 2011, Riviere et al. cooled a 70 MHz micromechanical silica oscillator to an occupancy around 9 quanta, implying that the mechanical oscillator can be found 10% of the time in its quantum ground state [49]. Chan et al. reported a nanoscale optical and mechanical resonator with frequency of 3.68 GHz formed in a photonic crystal cavity, in which radiation pressure from a laser is used to cool the mechanical motion down to its quantum ground state. The average phonon occupancy number is 0.85, and the cooling was realized at an environmental temperature of 20 K. In 2012, Verhagen et al. reported a spoke-anchored toroidal microcavity to achieve quantum-coherent coupling of a mechanical oscillator to an opitcal cavity mode, that is, the coherent coupling rate exceeded both the optical and the mechanical decoherence rate and quantum states were transferred from the optical field to the mechanical oscillator and vice versa [50]. The mechanical oscillator was cooled to an average occupancy of 1.7 motional quanta. In this regime, the system can be appropriately described as an optomechanical polariton whose decoherence time exceeded the period of the Rabi oscillations between light and mechanics.

In an optomechanical resonator, the optical and mechanical excitations can be coherently converted, leading to effects such as optomechanically induced transparency, which is the optomechanical analogy of the well-known phenomenon of electromagnetically induced transparency (EIT). In 2010, Weis et al. observed this phenomenon in a microtoroid cavity [51]. In 2011, Safavi-Naeini et al. demonstrated optomechanically induced transparency in a photonic crystal cavity, and reported an optically tunable delay of 50 ns with near-unity optical transparency and superluminal light with a 1.4 $\mu$s signal advance [52]. At the same time, Fiore et al. demonstrated the storing optical information as a mechanical excitation for several microseconds in a silica optomechanical resonator [53]. They used writing and readout laser pulses tuned to one mechanical frequency below the optical cavity resonance to control the coupling between the mechanical mode and the optical field at the cavity resonance. The writing pulse mapped a signal pulse at the cavity resonance to a mechanical excitation. The readout pulse later converted the mechanical excitation back to an optical pulse. The storage lifetime was determined by the damping time of the mechanical excitation. In 2012, Dong et al. realized a dark mode by coupling two optical modes in a silica resonator to one of its mechanical breathing modes in the regime of weak optomechanical coupling [54]. The dark mode was a superposition of the two optical modes and decoupled from the mechanical oscillator. On one hand, the formation of the dark mode enabled the transfer of optical fields between the two optical modes. On the other hand, the dark mode was decoupled from the mechanical oscillator and thus the mechanically mediated coupling can be robust against mechanical dissipation and thermal mechanical motion, which was important for a hybrid quantum network. In the experiment, the dark-mode fraction reached 99% and the conversion efficiency between photons of 637 and 800 nm could achieve more than 16%. Actually, the conversion between optical light and microwave could also be realized using optomechanical interactions [55]. In 2015, Cohen et al. used an optical probe and single-photon detection to study the acoustic emission and absorption processes in a photonic crystal and reported an effective phonon counting technique [56]. With straightforward improvements to the optomechanical technology in 2016, Riedinger et al. demonstrated nonclassical correlations between single photons and phonons, which were measured using the transduced photons in an optomechanical crystal cavity, and implemented a full quantum protocol for the generation of mechanical quantum states [57].

In the above experiments of cavity optomechanics, researchers usually focus on the breathing modes, where the mechanical oscillator expands and compresses uniformly. In fact, the whispering gallery microresonators also support the travelling acoustic wave. Similar to the optical field, the acoustic resonance travel around the circumference with an integer number of acoustical wavelengths and high quality factors. When the acoustic wave and two optical modes in the whispering gallery microresonators satisfy the energy and momentum conservations, photons can be scattered between the two optical resonances through Brillouin scattering. According to the scattering direction from forward to backward, the Brillouin scattering is also divided into forward and backward stimulated Brillouin scattering, as shown in Fig. 1.4. In recent years, researchers have observed stimulated Brillouin scattering in different types of optical microcavities. In 2009, Tomes et al. observed the optomechanical effects of backward stimulated Brillouin scattering in a dielectric microsphere cavity [40]. At the same year, Grudinin et al. used a 1064 nm laser to pump a calcium fluoride resonator with an ultrahigh quality factor to achieve Brillouin laser with a low threshold power of only 3 $\mu$W [58]. In 2011, Bahl et al. exploited a dielectric microsphere with a radius of around 100 microns to achieve forward stimulated Brillouin scattering at a frequency of 49–1400 MHz [59]. Subsequently, they observed both forward and back stimulated Brillouin scattering in a hollow optical resonator that can be filled with liquid, and the acoustic resonant frequency rangeing from 2 MHz to 11 GHz [60]. In 2012, they experimentally demonstrated cooling via a forward Brillouin process in a microresonator. The travelling acoustic mode with frequency of 95 MHz was cooled from room temperature to 19 K [61]. In 2013, Li et al. reported the application of cascaded Brillouin oscillation in an ultra-high-Q planar disk resonator [62].

It can be seen that Brillouin optomechanics has attracted great attention in recent years. Brillouin scattering in whispering-gallery microresonators is promising in several aspects. First, Brillouin scattering enables the optical coupling to acoustic waves with frequency ranging from a few MHz to 11 GHz, thereby providing a diverse platform for coherent light-matter interactions. Second, the triply resonant configuration can greatly enhance the Brillouin scattering, thereby reducing the power consumptions. In addition, the phase matching for the travelling waves enables non-reciprocal

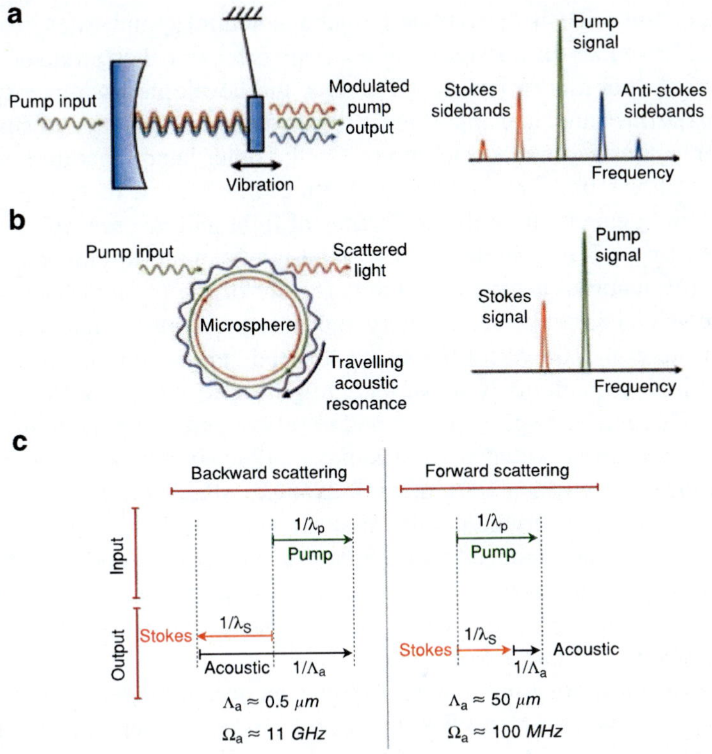


**Fig. 1.4** **a** The optomechanical interaction in the FP cavity. **b** The interaction between the travelling acoustic wave and optical whispering-gallery modes. **c** Forward and backward Brillouin scattering. Figure adapted from Ref. [60]

optical processes, thus offering potential application in an all optical integrated isolator and circulator devices. In 2015, our group and Bahl's group both demonstrated the Brillouin-scattering-induced transparency in a microcavity based on the forward travelling acoustic wave [63, 64]. Furthermore, we used this light-acoustic interaction to achieve the storage of light as a coherent, circulating acoustic wave with a lifetime up to 10 $\mu$s [64]. Because of the phase-matching requirement, a circulating acoustic wave can only couple to light with a given propagation direction, leading to non-reciprocal light storage and retrieval.

Optomechanical interactions are also studied in optical waveguides, such as photonic circuits or photonic crystal fibres in the absence of a cavity [65, 66]. They are very promising for applications due to their large bandwidth. In addition, researchers successfully realized optomechanical coupling in superconducting cavities by embedding a nanomechanical beam inside a superconducting transmission line microwave cavity or by incorporating a flexible aluminum membrane into a lumped element superconducting resonator [67, 68]. Their capacitive elements as high-quality mechanical modes coupled to the microwave cavity. The combination of optomechanics and micro-electro-mechanical system (MEMS) that has been garnering momentum since the late 1990s, has great potential applications in the sensing field [69], and provides the possibility of realizing mechanical quantum devices [70].

There are several motivations to drive the rapid development of cavity optomechanics. On the one hand, it enables highly sensitive optical detection of small forces, displacements, masses, and accelerations. On the other hand, quantum cavity photomechanics allows us to manipulate and detect mechanical motion in the quantum regime and generate non-classical states of light and macroscopic mechanical devices, thereby studying the decoherence and the quantum-to-classical transition. For the hybrid quantum network, a unique feature of the optomechanical systems is the interconversion between stationary and flying (photonic) qubits. Mechanical motion can serve as a universal transducer to mediate the long-range interactions between stationary quantum systems, including trapped ions, superconducting circuits, single charges, and spins in diamond or silicon, enabling the construction of a hybrid quantum network that combines the otherwise incompatible degrees of the freedom of different physical systems. For example, the quantum information that was processed by the superconducting circuits can be stored in mechanical motion via the optomechanically induced transparency or can be converted into photons that were transferred to other distant quantum nodes through optical fibers. During the past few years, a number of reviews on cavity optomechanics have been published [71–74], showing great interest of researchers.

As a carrier for studying cavity optomechanics, whispering-gallery-modes microresonators have attracted wide attention from researchers due to their ultra-high quality factor (Q) and small mode volume (V). They are not only one of the important experimental platforms in the field of cavity optomechanics, but also can be used to realizes ultra-low threshold lasers [58, 75, 76], parametric oscillator [77], narrowband optical filters [78], highly sensitive weak force and biological sensor [79–81], cavity quantum electrodynamic with atoms [82, 83] and optical frequency comb [84, 85].

Our group was established in 2013. Research in this group is centered around ultra-high-Q optical microresonators, including both theory and experiment. In 2016, our group experimentally demonstrated non-magnetic nonreciprocity using optomechanical interactions in a whispering-gallery microresonator. The mechanical mode can also be more general, such as the breathing modes of the resonators and the external nanomechanical oscillators coupled to the resonators. Therefore, The underlying mechanism of optomechanically induced non-reciprocity has great potential for all-optical controllable isolators and circulators, as well as non-reciprocal phase shifters in integrated photonic chips. In addition, the optomechanical interaction between the travelling optical and mechanical whispering-gallery modes will be introduced in this thesis. We demonstrate the Brillouin-scattering induced transparency in a high-quality whispering-gallery-mode optical microresonantor. The triply resonant Stimulated Brillouin scattering process underlying the Brillouin scattering-induced transparency greatly enhances the light-acoustic interaction, enabling the storage of light as a coherent, circulating acoustic wave. In the last chapter, we will introduce the technology of phase sensitive imaging of mechanical vibration, which can give us an intuitive understanding of cavity optomechanics.

## References

1. Maxwell JC (1881) A treatise on electricity and magnetism, vol 1. Clarendon Press, Oxford
2. Lebedew P (1901) Annalen der Physik 311:433
3. Nichols EF, Hull GF (1901) Phys Rev 13:307
4. Hansch TW, Schawlow AL (1975) Opt Commun 13:68
5. Wineland DJ, Itano WM (1979) Phys Rev A 20:1521
6. Ashkin A (1978) Phys Rev Lett 40:729
7. Chu S, Hollberg L, Bjorkholm JE, Cable A, Ashkin A (1985) Phys Rev Lett 55(1):48
8. Stenholm S (1986) Rev Mod Phys 58:699
9. Metcalf HJ, van der Straten P (1999) Laser cooling and trapping. Springer, New York
10. Bloch I, Dalibard J, Zwerger W (2008) Rev Mod Phys 80:885
11. Diedrich F, Bergquist JC, Itano WM, Wineland DJ (1989) Phys Rev Lett 62(4):403
12. Hamann SE, Haycock DL, Klose G, Pax PH, Deutsch IH, Jessen PS (1998) Phys Rev Lett 80(19):4149
13. Leibfried D, Blatt R, Monroe C, Wineland D (2003) Rev Mod Phys 75(1):281
14. Braginski VB, Manukin AB (1967) Sov Phys JETP 25(4):653
15. Braginski VB, Manukin AB, Tikhonov MY (1970) Sov Phys JETP 31:829
16. Dorsel A, McCullen JD, Meystre P, Vignes E, Walther H (1983) Phys Rev Lett 51:1550
17. Braginski VB, Khalili FYA (1977) Measurement of weak forces in physics experiments. Cambridge University Press, Chicago
18. Braginski VB, Manukin AB (1995) Quantum measurements. University of Chicago Press, Cambridge
19. Caves CM (1980) Phys Rev Lett 45:75
20. Jaekel M, Renaud S (1991) J Phys I (France) 1(10):1395
21. Jacobs K, Tombesi P, Collett MJ, Walls DF (1994) Phys Rev A 49(3):1961
22. Fabre C, Pinard M, Bourzeix S, Heidmann A, Giacobino E, Reynaud S (1994) Phys Rev A 49(2):1337
23. Bose S, Jacobs K, Knight PL (1997) Phys Rev A 56(5):4175
24. Hechenblaikner G, Gangl M, Ritsch H (1998) Phys Rev A 58(4):3030
25. Cohadon PF, Heidmann A, Pinard M (1999) Phys Rev Lett 83(16):3174
26. Tittonen I, Breitenbach G, Kalkbrenner T, Muller T, Conradt R, Steinsland SSE, Blanc N, de Rooij NF (1999) Phys Rev A 59(2):1038
27. Vogel M, Mooser C, Karrai K, Warburton RJ (2003) Appl Phys Lett 83(7):1337
28. Mertz J, Marti O, Mlynek J (1993) Appl Phys Lett 62:2344
29. Zalalutdinov M, Zehnder A, Olkhovets A, Turner S, Sekaric L, Ilic B, Czaplewski D, Parpia JM, Craighead HG (2001) Appl Phys Lett 79:695
30. Metzger CH, Karrai K (2004) Nature (London) 432:1002
31. Carmon T, Rokhsari H, Yang L, Kippenberg TJ, Vahala KJ (2005) Phys Rev Lett 94(22):223902
32. Kippenberg TJ, Rokhsari H, Carmon T, Scherer A, Vahala KJ (2005) Phys Rev Lett 95(3):033901
33. Rokhsari H, Kippenberg TJ, Carmon T, Vahala KJ (2005) Opt Express 13(14):5293
34. Thompson JD, Zwickl BM, Jayich AM, Marquardt F, Girvin SM, Harris JGE (2008) Nature (London) 452:72
35. Favero I, Stapfner S, Hunger D, Paulitschke P, Reichel J, Lorenz H, Weig EM, Karrai K (2009) Opt Express 17(15):12813
36. Jiang X, Lin Q, Rosenberg J, Vahala K, Painter O (2009) Opt Express 17(23):20911
37. Wiederhecker GS, Chen L, Gondarenko A, Lipson M (2009) Nature (London) 462:633
38. Ma R, Schliesser A, Del'Haye P, Dabirian A, Anetsberger G, Kippenberg TJ (2007) Opt Lett 32(15):2200
39. Park YS, Wang H (2009) Nat Phys 5:489
40. Tomes M, Carmon T (2009) Phys Rev Lett 102(11):113601
41. Eichenfield M, Camacho R, Chan J, Vahala KJ, Painter O (2009) Nature (London) 459:550

42. Eichenfield M, Chan J, Camacho RM, Vahala KJ, Painter O (2009) Nature (London) 462:78
43. Anetsberger G, Arcizet O, Unterreithmeier QP, Riviere R, Schliesser A, Weig EM, Kotthaus JP, Kippenberg TJ (2009) Nat Phys 5:909
44. Arcizet O, Cohadon PF, Briant T, Pinard M, Heidmann A (2006) Nature (London) 444:71
45. Gigan S, Bohm HR, Paternostro M, Blaser F, Langer G, Hertzberg JB, Schwab KC, Anspelmeyer DB, Zeilinger A (2006) Nature (London) 444:67
46. Schliesser A, Del'Haye P, Nooshi N, Vahala KJ, Kippenberg TJ (2006) Phys Rev Lett 97(24):243905
47. Schliesser A, Riviere R, Anetsberger G, Arcizet O, Kippenberg T (2008) Nat Phys 4:415
48. Schliesser A, Arcizet O, Riviere R, Kippenberg T (2009) Nat Phys 5:509
49. Riviere R, Deleglise S, Weis S, Gavartin E, Arcizet O, Schliesser A, Kippenberg TJ (2011) Phys Rev A 83(6):063835
50. Verhagen E, Deleglise S, Weis S, Schliesser A, Kippenberg TJ (2012) Nature 482:63
51. Weis S, Riviere R, Deleglise S, Gavartin E, Arcizet O, Schliesser A, Kippenberg TJ (2010) Science 330(6010):1520
52. Safavi-Naeini AH, Alegre TPM, Chan J, Eichenfield M, Winger M, Lin Q, Hill JT, Chang DE, Painter O (2011) Nature 472:69
53. Fiore V, Yang Y, Kuzyk MC, Barbour R, Tian L, Wang H (2011) Phys Rev Lett 107(13):133601
54. Dong C, Fiore V, Kuzyk MC, Wang H (2012) Science 338(6114):1609
55. Bagci T, Simonsen A, Schmid S, Villanueva LG, Zeuthen E, Appel J, Taylor JM, Sorensen AS, Usami K, Schliesser A et al (2014) Nature 507:81
56. Cohen JD, Meenehan SM, MacCabe GS, Safavi-Naeini AH, Marsili F, Shaw MD, Painter O et al (2015) Nature 520:522
57. Riedinger R, Hong S, Norte RA, Slater JA, Shang J, Krause AG, Anant V, Aspelmeyer M, Groblacher S (2016) Nature 530:313
58. Grudinin IS, Matsko AB, Maleki L (2009) Phys Rev Lett 102(4):043902
59. Bahl G, Zehnpfennig J, Tomes M, Carmon T (2011) Nat Commun 2
60. Bahl G, Kim KH, Lee W, Liu J, Fan X, Carmon T (2013) Nat Commun 4
61. Bahl G, Tomes M, Marquardt F, Carmon T (2012) Nat Phys 8:203
62. Li J, Lee H, Vahala KJ (2013) Nat Commun 4
63. Kim J, Kuzyk MC, Han K, Wang H, Bahl G (2015) Nat Phys 11:275
64. Dong CH, Shen Z, Zou CL, Zhang YL, Fu W, Guo GC (2015) Nat Commun 6:6193
65. Li M, Pernice WHP, Xiong C, Baehr-Jones T, Hochberg M, Tang HX (2008) Nature (London) 456:480
66. Kang MS, Nazarkin A, Brenn A, Russell PSJ (2009) Nat Phys 5:276
67. Regal CA, Teufel JD, Lehnert KW (2008) Nat Phys 4:555
68. Teufel JD, Li D, Allman MS, Cicak K, Sirois AJ, Whittaker JD, Simmonds RW (2011) Nature (London) 471:204
69. Cleland AN, Roukes ML (1998) Nature (London) 392:160
70. Blencowe MP (2005) Contemp Phys 46(4):249
71. Kippenberg T, Vahala K (2007) Opt Express 15(25):17172
72. Marquardt F, Girvin SM (2009) Physics 2:40
73. Aspelmeyer M, Meystre P, Schwab K (2012) Phys Today 65(7):29
74. Aspelmeyer M, Kippenberg TJ, Marquardt F (2014) Rev Mod Phys 86(4):1391
75. Spillane S, Kippenberg T, Vahala K (2002) Nature 415(6872):621
76. He L, Ozdemir SK, Yang L (2013) Laser Photonics Rev 7(1):60
77. Furst JU, Strekalov DV, Elser D, Aiello A, Andersen UL, Marquardt C, Leuchs G (2010) Phys Rev Lett 105(26):263904
78. Savchenkov AA, Ilchenko VS, Matsko AB, Maleki L (2005) IEEE Photonics Technol Lett 17(1):136
79. Armani AM, Kulkarni RP, Fraser SE, Flagan RC, Vahala KJ (2007) Science 317(5839):783
80. Gavartin E, Verlot P, Kippenberg TJ (2012) Nat Nanotechnol 7(8):509
81. He L, Ozdemir SK, Zhu J, Kim W, Yang L (2011) Nat Nanotechnol 6(7):428
82. Dayan B, Parkins A, Aoki T, Ostby E, Vahala K, Kimble H (2008) Science 319(5866):1062

83. Boca A, Miller R, Birnbaum K, Boozer A, McKeever J, Kimble H (2004) Phys Rev Lett 93(23):233603
84. Papp SB, Beha K, Del'Haye P, Quinlan F, Lee H, Vahala KJ, Diddams SA (2014) Optica 1(1):10
85. Brasch V, Geiselmann M, Herr T, Lihachev G, Pfeiffer MH, Gorodetsky ML, Kippenberg TJ (2016) Science 351(6271):357

# Chapter 2
# Whispering Gallery Modes Microcavity

**Abstract** This chapter introduces the origin of the Whispering-gallery modes (WGM) microcavities. The WGM microcavities have independent optical and mechanical properties. Several terminologies, such as optical or mechanical quality factor, used in cavities are introduced. The mode characterization and field distribution of the microsphere is discussed. This chapter also describes the fabrication process of microresonator and tapered fiber, and gives a review of the tapered fiber coupling schemes used in our work. A simple theoretical model is presented to study the coupling mechanism between the microcavities and tapered fiber.

## 2.1 Optical Whispering Gallery Modes

The term whispering gallery modes (WGMs) was derived from the acoustic phenomenon studied by Lord Rayleigh. Under the dome with a diameter of 32 m in St. Paul's Cathedral in London, if one speaks quietly against the stone wall, people standing on the other side of the gallery can also hear clearly [1], as shown in Fig. 2.1a, b. Rayleigh explained the phenomenon as sound waves could travel around the walls of the dome. The reflection and refraction will occur when light rays hit the interface between two media. Total internal reflection takes place at the boundary between two transparent media when a ray of light in a medium of higher index of refraction approaches the other medium at an angle of incidence greater than the critical angle. For circular dielectric microcavities, such as microspheres or microdisks, light rays within the cavities can bounce off the surface due to total internal reflection. If the path length of the circumference is an integer multiple of the light wavelength, a resonant optical mode can be formed, called the whispering gallery mode, as shown in Fig. 2.1c, d. In this chapter we will introduce the whispering gallery modes [2–5].



Z. Shen, *Experimental Research of Cavity Optomechanics*, Springer Theses,
https://doi.org/10.1007/978-981-33-4458-7_2

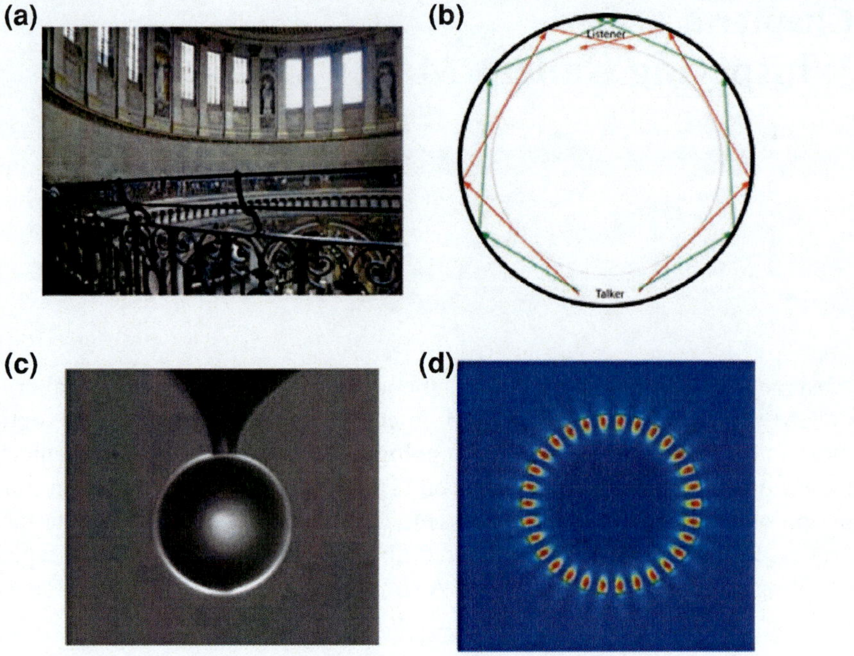


**Fig. 2.1** **a** The whispering gallery of St. Paul's Cathedral. **b** Schematic of the whispering gallery. **c** A photograph of the silica microsphere. **d** Optical whispering gallery mode

### *2.1.1 Electromagnetic Field in a Spherical Cavity*

To investigate the optical resonance modes of a dielectric sphere of radius $R$ and refractive index $n$, we study the Maxwell's equation in an isotropic medium with free of charge and current:

$$\nabla \cdot \mathbf{D} = 0 \tag{2.1}$$

$$\nabla \cdot \mathbf{B} = 0 \tag{2.2}$$

$$\nabla \times \mathbf{E} = -\frac{\partial \mathbf{B}}{\partial t} = -\mu \frac{\partial \mathbf{H}}{\partial t} \tag{2.3}$$

$$\nabla \times \mathbf{H} = -\frac{\partial \mathbf{D}}{\partial t} = \varepsilon \frac{\partial \mathbf{E}}{\partial t} \tag{2.4}$$

We can use the vector identity:

$$\nabla \times \nabla \times \mathbf{A} = \nabla(\nabla \cdot \mathbf{A}) - \nabla^2 \mathbf{A} \tag{2.5}$$

to obtain:

$$\nabla(\nabla \cdot \mathbf{E}) - \nabla^2 \mathbf{E} = -\mu \nabla \times \frac{\partial \mathbf{H}}{\partial t} \tag{2.6}$$

According to Maxwell's equations, we can simplify the above equation as:

$$\nabla^2 \mathbf{E} - \mu \varepsilon \frac{\partial^2 \mathbf{E}}{\partial t^2} = 0 \tag{2.7}$$

After separating the spatial and temporal variations of the electric field, $\mathbf{E}(\mathbf{r}, t) = E(\mathbf{r})e^{i\omega t}$, we obtain the vectorial Helmholtz equation:

$$\triangle \mathbf{E}(\mathbf{r}) + k(r)^2 \cdot \mathbf{E}(\mathbf{r}) = 0 \tag{2.8}$$

where $\mathbf{E}(\mathbf{r})$ is the electric field, $k$ is wavenumber, for inside the sphere $k(r) = nk_0$, and outside the sphere $k(r) = k_0$. In a homogeneous medium, the electric field can be calculated analytically using the Debye's potential. The equation has two classes of solutions, TE modes and TM modes. First, the electric field is separated into radial and orthoradial variations, and then by solving the scalar Helmholtz equation in the spherical coordinate system, the electric field of TE mode and TM mode can be obtained:

$$\mathbf{E}_{l,m}^{TE}(r, \theta, \varphi) = E_0 \frac{f_l(r)}{k_0 r} \mathbf{Y}_l^m(\theta, \varphi) \tag{2.9}$$

$$\mathbf{B}_{l,m}^{TE}(r, \theta, \varphi) = -\frac{i}{ck_0} \nabla \times \mathbf{Y}_l^m(\theta, \varphi) \tag{2.10}$$

and

$$\mathbf{E}_{l,m}^{TM}(r, \theta, \varphi) = -\frac{i}{ck_0 n^2} \nabla \times \mathbf{B}_{l,m}^{TM}(\theta, \varphi) \tag{2.11}$$

$$\mathbf{B}_{l,m}^{TM}(r, \theta, \varphi) = -\frac{iE_0}{c} \frac{f_l(r)}{k_0 r} \mathbf{Y}_l^m(\theta, \varphi) \tag{2.12}$$

where $\mathbf{Y}_l^m(\theta, \varphi)$ is the vector spherical harmonic function:

$$\mathbf{Y}_l^m(\theta, \varphi) = \frac{1}{\sqrt{l(l+1)}} \nabla Y_l^m(\theta, \varphi) \times \mathbf{r} \tag{2.13}$$

which can be derived from the standard scalar spherical harmonic function:

$$Y_l^m(\theta, \varphi) = \sqrt{\frac{2l+}{4\pi} \frac{(l-m)!}{(l+m)!}} P_l^m(cos(\theta)) e^{im\varphi} \tag{2.14}$$

where $P_l^m$ is Legendre polynomial. The radial variation is described by the solution of the Riccati–Bessel function:

$$f_l(r) = \begin{cases} nk_0 r j_l(nk_0 r) & = \sqrt{\frac{\pi n k_0 r}{2}} J_{l+1/2}(nk_0 r) \quad \text{for } r < R \\ \alpha k_0 r h_l(k_0 r) & = \alpha \sqrt{\frac{\pi k_0 r}{2}} H_{l+1/2}(k_0 r) \quad \text{for } r > R \end{cases} \tag{2.15}$$

where $j_l$ and $h_l$ ($J_l$ and $H_l$) are the spherical (cylindrical) Bessel and Hankel functions of the first kind, respectively and $\alpha$ is a real constant.

The continuity conditions of the electromagnetic field at the boundary require:

$$nk_0 R j_l(nk_0 R) = \alpha k_0 R h_l(k_0 R) \tag{2.16}$$

$$n^s [nk_0 R j_l(nk_0 R)]' = \alpha [k_0 R h_l(k_0 R)]' \tag{2.17}$$

where $s = +1$ for TE mode and $s = -1$ for TM mode. These characteristic equations determine not only the constant $\alpha$, but also the resonant wavenumber $k_0$ and frequency $\omega_c = ck_0$ of the whispering gallery mode.

For different $l$ value, due to the nature of Bessel and Hankel functions, the Eq. (2.17) supports an infinite number of solutions. These solutions can be distinguished by the parameter $q$, which determines the number of nodes in the radial direction of the electric field in the sphere. It is usually called the radial mode number, which is different from the polar mode number $l$ and the azimuthal mode number $m \in \{-l, \ldots, +l\}$. The most common optical WGMs in experiments usually have low $q$, high $l$, and $|m| \sim l$. It should be pointed out that for a perfect sphere, the resonance frequency of the WGM is independent of $m$, because the characteristic equation is independent of $m$. However, in the fabrication process of the microspheres, due to the gravity and other environmental fluctuations, the deformed microsphere cavities lift this degeneracy. Only the two modes that propagate in opposite directions, $m$ and $-m$, remain degenerate. Note that the electromagnetic field directions of TE mode and TM mode are almost orthogonal. For TE mode, the electric field is essentially parallel to the polar direction (along the unit vector $\mathbf{e}_\theta$). For TM mode, the magnetic field is essentially parallel to the polar direction.

We can obtain the resonant frequencies of the WGMs with some effective approximations to the characteristic equations for large $l$. The resonance frequency of the fundamental mode is:

$$\nu_l = \frac{c}{2\pi n R} \left( l + 1/2 + \eta_1 \left( \frac{l + 1/2}{2} \right)^{1/3} + \cdots \right) \tag{2.18}$$

where $c$ is the speed of light in vacuum, $n$ is the refractive index, $R$ is the radius of the spherical cavity, $-\eta_1$ is the first zero of the Airy equation ($-\eta_1 \approx 2.34$).

For the microdisk cavities, the TE mode and TM mode solutions can also be obtained by solving the Helmholtz equation (2.8) in the cylindrical coordinates. The solution process is similar to that of the microsphere cavities and can be referred to Ref. [6].

### 2.1.2 Optical Loss

A lossless optical microcavity can store light for any length of time. However, due to the various loss mechanisms, the energy stored in a practical cavity will decay on the time scale $\tau$, or at the loss rate $\kappa$,

$$\kappa = 1/\tau \tag{2.19}$$

Photon lifetime $\tau$ and loss rate $\kappa$ are often described by the quality factor $Q$ of the microcavity, which is important for practical applications, such as microcavity lasers, cavity quantum electrodynamics, and cavity optomechanics. It is defined as the ratio between the energy stored in the cavity and the energy lost during a oscillation cycle:

$$Q \equiv \tau\omega_c = \omega_c/\kappa \tag{2.20}$$

where $\omega_c$ is the resonant frequency of the WGM.

For the silica microcavity, there are multiple dissipation mechanisms that will reduce the photon lifetime $\tau$, for example, the absorption of silica materials and surface attachments, radiative loss and light scattering by the inhomogeneities in bulk or at surface. In addition, the coupling of WGM and the propagation mode in the optical waveguide will also induce losses. These losses are usually quantified by the rate $\tau_{mat}^{-1}$, $\tau_{con}^{-1}$, $\tau_{rad}^{-1}$, $\tau_{sca}^{-1}$, $\tau_{ex}^{-1}$, and the total loss rate is:

$$\kappa = \tau^{-1} = \tau_{mat}^{-1} + \tau_{con}^{-1} + \tau_{rad}^{-1} + \tau_{sca}^{-1} + \tau_{ex}^{-1} \tag{2.21}$$

The intrinsic loss usually refers to:

$$\tau_0^{-1} = \tau_{mat}^{-1} + \tau_{con}^{-1} + \tau_{rad}^{-1} + \tau_{sca}^{-1} \tag{2.22}$$

The corresponding intrinsic quality factor is defined as:

$$Q_0^{-1} = (\omega_c\tau_0)^{-1} = Q_{mat}^{-1} + Q_{con}^{-1} + Q_{rad}^{-1} + Q_{sca}^{-1} \tag{2.23}$$

The pure silica is transparent in the visible and near-infrared bands. The upper limit of the quality factor determined by the material absorption can be simply expressed as $Q_{mat} = 2\pi n/\alpha\lambda$, where $\alpha$ is the absorption coefficient determined by the material absorption and Rayleigh scattering. The theoretical limit of the $Q$ factor can achieve $10^{10}$ at 633 nm, and experimental results close to this limit are reported in the Ref. [7]. For longer wavelengths, the $Q_{mat}$ can even exceed $10^{11}$ [8].

The radiation loss originates from the curved surface of the microcavity that is unable to adequately confinement the light field, leading the energy lost. For most of the silica microsphere cavities used in this thesis, the radiation loss can be ignored. In theory, the quality factor dominated by radiation loss exceeds $10^{11}$ at both the visible and near-infrared bands for microspheres with a diameter larger than 20 μm [9].

Another loss mechanism is the light scattering by the inhomogeneities of microcavity surface [8]. Although surface tension is exploited in the fabrication process of silica microspheres, giving rise to the smooth surface. The redeposition of evaporated silica will cause the surface defects at the sub-wavelength scale. The $Q$ dominated by this scattering can be written as $Q_{sca} \propto \lambda^3/\sigma^2 B^2$, where $\sigma$ is the surface roughness, and $B$ is correlation length [10]. The scattering loss also depends on the field distribution of the optical mode. The modes where the light field tends to be distributed on the surface will suffer more losses. For the microsphere cavity with diameter of 750 μm, the $Q$ factor can reach $8 \times 10^9$ [7], which is close to the theoretical upper limit of the quality factor determined by scattering loss $Q_{sca}$ [8].

In most experiments performed under ambient conditions, the main loss comes from the adsorption of water attached to the surface of silica material [7, 10, 11]. In particular at 1.5 μm band used in communication applications, water will absorb photons strongly, even a monolayer of water molecules will cause significant loss, resulting in the highest quality factor of $3 \times 10^8$ for a 60 μm microsphere [11]. The reduction of $Q_{con}$ occurs within 100 s after preparation of the microcavity and can be partially recovered by baking at 400 °C [7].

The above content focuses mainly on the dissipation mechanisms of silica microsphere cavities. These dissipation mechanisms also contribute to the microspheres, microdisks and microtoroids of other materials.

## 2.2 Mechanical Modes

The optomechanical microcavities always support a series of mechanical normal modes with different resonant frequencies, as shown in Fig. 2.2. Here, we only focus on a single mechanical mode of vibration of frequency $\omega_m$ and mechanical energy

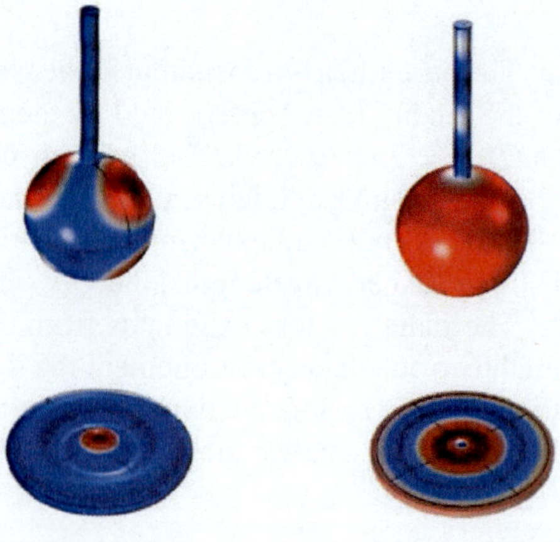

**Fig. 2.2** Typical mechanical modes in microsphere and microdisk cavities

damping rate $\omega_m$. We can use the quality factor $Q_m = \omega_m/\gamma_m$ to describe the mechanical quality related to the energy loss. If the displacement field of mechanical vibration is written as $\mathbf{u}(\mathbf{r}, t) = x(t) \cdot \mathbf{u}(\mathbf{r})$, then $x(t)$ follows the simple equation of motion of a harmonic oscillator:

$$m_{eff}\frac{dx^2(t)}{dt^2} + m_{eff}\gamma_m\frac{dx(t)}{dt} + m_{eff}\omega_m^2 x(t) = F_{ext}(t) \tag{2.24}$$

where $F_{ext}$ represents the external force, and $m_{eff}$ is effective mass. This equation can be easily solved by Fourier transform in frequency domain as $x(\omega) = \chi(\omega)F_{ext}(\omega)$, where $\chi$ is the response function of the oscillator:

$$\chi(\omega) = [m_{eff}(\omega_m^2 - \omega^2) - im_{eff}\gamma_m\omega] \tag{2.25}$$

The mechanical harmonic oscillator can be quantized and its Hamiltonian is expressed as:

$$\mathbf{H} = \hbar\omega_m\hat{b}^\dagger\hat{b} + \frac{1}{2}\hbar\omega_m \tag{2.26}$$

$\hat{b}^\dagger$ and $\hat{b}$ are phonon creation and annihilation operators, corresponding to

$$\hat{x} = x_{ZPF}(\hat{b} + \hat{b}^\dagger) \tag{2.27}$$

$$\hat{p} = -im_{eff}\omega_m x_{ZPF}(\hat{b} - \hat{b}^\dagger) \tag{2.28}$$

where

$$x_{ZPF} = \sqrt{\frac{\hbar}{2m_{eff}\omega_m}} \tag{2.29}$$

is the zero fluctuation amplitude of the mechanical oscillator, that is, $\langle 0 \left| \hat{x}^2 \right| 0 \rangle = x_{ZPF}^2$, and where $|0\rangle$ represents a mechanical vibration vacuum state. The displacement and momentum satisfy the commutator relation $[\hat{x}_{ZPF}, \hat{p}_{ZPF}] = i\hbar$, and average phonon number is denoted by $n = \left\langle \hat{b}^\dagger\hat{b} \right\rangle$.

When a mechanical oscillator is coupled to a high-temperature thermal reservoir, the average number of phonons will evolve as follows:

$$\frac{d}{dt}\langle n \rangle = -\gamma_m(\langle n \rangle - \bar{n}_{th}) \tag{2.30}$$

For a harmonic oscillator which is initially in the ground state, the number of phonons is $\langle n \rangle (t) = \bar{n}_{th}(1 - e^{-t/\gamma_m})$ versus time $t$, where $\bar{n}_{th}$ is the average number of the environment. Therefore, the thermal decoherence rate of the mechanical oscillator is given by:

$$\frac{d}{dt}\langle n \rangle (t = 0) = \bar{n}_{th} \cdot \gamma_m \approx \frac{k_B T}{\hbar Q_m} \tag{2.31}$$

The above expression shows that high mechanical quality factor and low temperature are important to achieve low thermal decoherence.

The mechanisms affecting the $Q$ factor of the mechanical mode generally have the following aspects [12–14]:

Viscous Damping

It comes from the interaction of the mechanical oscillator with the molecules in the surrounding gas or liquid environment.

Clamping Losses

Due to the radiation of elastic waves into the substrate through the support structure of the mechanical oscillators.

Fundamental Anharmonic Effects

Caused by the thermoelastic effect and phonon-phonon interactions.

Materials-Induced Losses
Due to the inherent defects in the bulk or surface of the material.

Similar to the dissipation of optical mode, the total loss rate of the mechanical mode is the sum of various dissipation processes.

Due to the effects of mechanical damping and fluctuating thermal Langevin force, the amplitude and phase of the mechanical vibration change randomly on the time scale determined by the damping time $\gamma_m^{-1}$.

Although the amplitude and phase can be measured in real-time in optomechanical systems. In general experiments, we usually observe the mechanical motion by measuring the noise spectrum in the frequency domain. This method also allows us to easily distinguish different mechanical normal modes. For the mechanical vibration displacement $x(t)$ measured in a finite duration $\tau$ if the measurement time is a finite duration $\tau$, the gated Fourier transform is:

$$\tilde{x}(\omega) = \frac{1}{\sqrt{\tau}} \int_0^{\tau} x(t) e^{i\omega t} dt \tag{2.32}$$

Ensemble averaging the measurement, we can get the spectral density $\left\langle |\tilde{x}(\omega)|^2 \right\rangle$. In the limit $\tau \to \infty$, it is equivalent to the Fourier transform of the autocorrelation function based on the Wiener–Khinchin theorem:

$$\left\langle |\tilde{x}(\omega)|^2 \right\rangle = S(\omega) \equiv \int_{-\infty}^{+\infty} \langle x(t)x(0) \rangle \, e^{i\omega t} dt \tag{2.33}$$

Therefore, the variance of mechanical displacement $\langle x^2 \rangle$ can be obtained by experimentally integrating the measured mechanical noise spectrum:

$$\int_{-\infty}^{+\infty} S(\omega)\frac{d\omega}{2\pi} = \langle x^2 \rangle \tag{2.34}$$

In thermal equilibrium, the noise power spectral density corresponds to the dissipative part of the response function:

$$S(\omega) = 2\frac{k_B T}{\omega}\mathrm{Im}\chi(\omega) \tag{2.35}$$

For weak damping $\gamma_m \ll \omega_m$, the noise power spectrum will have Lorentzian peaks of width $\gamma_m$ at $\omega = \pm\omega_m$, and the area corresponds to the variance of mechanical motion, which is governed by the equipartition theorem $\langle x^2 \rangle = k_B T / m_{eff}\omega_m^2$.

## 2.3 Microsphere Fabrication

In the experiment carried out in this thesis, silica microspheres are fabricated by melting a tapered fiber with a $CO_2$ laser, as shown in Fig. 2.3. First, a short section of optical fiber is clamped by a copper rod and a small weight attached to the bottom of the fiber. A ZnSe lens with a focal length of 3.8 cm is used to focus the laser on the tip of the fiber. The fiber is melted and stretched by the small weight suspended. After retaining the required length, the rest is cut off by the laser. Then, the heating is started from the lowest end and the laser spot gradually moved up. The melted optical fiber will naturally form a microsphere cavity with a smooth surface due to the effect of surface tension. The whole process is monitored in real time by a CCD. For the study of cavity optomechanics, the size control of the stem-microsphere system is important. The diameter and shape of the stem have a great influence on the quality factor of the mechanical mode.

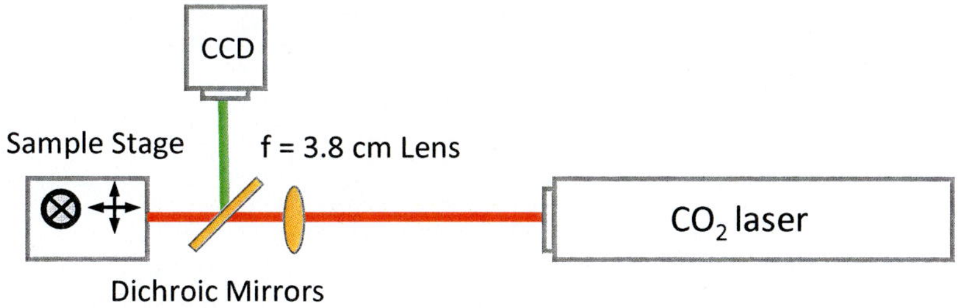


**Fig. 2.3** Setup for fabricating the microsphere cavity

## 2.4 Tapered Fiber Coupling

The excitation of the WGMs of microsphere cavity is mainly achieved by the tapered fiber. The coupling between the optical waveguide and WGMs via the overlap of their respective evanescent field enables the energy into and out of the microsphere cavity.

### *2.4.1 Tapered Fiber Fabrication*

The tapered fiber waveguide is obtained by melting a standard SMF-28 silica fiber using a hydrogen flame. The diameter of the thinnest part is on the order of light wavelength, which can be regarded as a silica cylindrical waveguide. Due to the difference of refractive index between the silica waveguide and surrounding air, the light field can be confined. When the diameter $2R$ of the cylindrical dielectric waveguide small enough, only a single propagating mode is supported. Along the radial direction of waveguide, the field intensity of the propagation mode decays exponentially with a characteristic length of $\gamma_f^{-1}$,

$$\gamma_f = \alpha_f \frac{K_1(\alpha_f R)}{K_0(\alpha_f R)} \tag{2.36}$$

where $K_0$, $K_1$ are zero- and first-order Hankel functions, respectively, and

$$\alpha_f = \sqrt{\beta_f^2 - k^2 n_s^2} \tag{2.37}$$

obtained from the characteristic equation:

$$k_f \frac{J_1(k_f R)}{J_0(k_f R)} = \alpha_f \frac{K_1(\alpha_f R)}{K_0(\alpha_f R)} \tag{2.38}$$

where $J_0$, $J_1$ are the zero- and first-order cylindrical Bessel equations, respectively. According to the above equation, we find:

$$k_f = \sqrt{k^2 n_f^2 - \beta_f^2} \tag{2.39}$$

where $\beta_f$ is the propagation constant, $n_f$ and $n_s$ are the refractive indices of the tapered fiber and the surrounding material, respectively. In the experiment, we adjusted the relative position of the tapered fiber and the microsphere cavity to achieve the purpose of changing the fiber radius. By continuously tuning the propagation constant of the tapered fiber, the phase matching with a specified WGM can be achieved, thereby achieving efficient coupling [15].

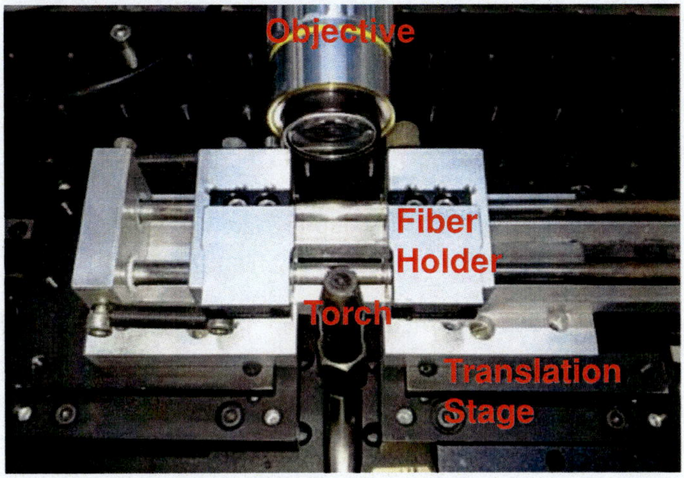


**Fig. 2.4** Setup used to prepare tapered fibers

The setup used to prepare tapered fibers is shown in Fig. 2.4. The single-mode fiber of the desired wavelength is clamped above the oxyhydrogen torch and melted using the hydrogen flame fed by hydrogen generator at a controlled flow. At the same time, two translation stages symmetrically pull the clamps apart at a uniform speed of 0.08 mm/s. During the pulling process, the tapered fiber is monitored in real time, and the transmission spectrum is detected by an optical power meter. When the fiber is pulled to a critical size, the transmission will oscillate due to multimode interference, that is, the transmittance depends on the length of the tapered fiber and pulling time. When the multimode interference disappears and the tapered fiber supports only a single mode, stop pulling and turn off the hydrogen flame, indicating the preparation of tapered fibers is completed.

For the preparation of tapered fibers, the acrylate buffer as a protection layer of the fiber should be completely removed before heating by mechanical stripping and wiping with a dust-free paper and ethanol. Fine adjustment of the relative position of the fiber and the hydrogen flame and the flux of the hydrogen generator, to ensure that the electronically controlled translation stage is stable, and to isolate the surrounding air currents, these are all important for the fabrication of low-loss tapered fiber.

## 2.4.2 *Input-Output Theory*

For a microcavity coupled with the outside electromagnetic environment, the quantum mechanical description can be obtained via the input-output theory. The intracavity field evolves over time according to the dynamic equation as:

$$\dot{\hat{a}} = i\Delta\hat{a} - \frac{\kappa}{2}\hat{a} + \sqrt{\kappa_{ext}}\hat{a}_{in} \tag{2.40}$$

The transmitted field is written as:

$$\hat{a}_{out} = \hat{a}_{in} - \sqrt{\kappa_{ext}}\hat{a} \tag{2.41}$$

where $\hat{a}$ is the annihilation operator of the optical mode, the detuning $\Delta = \omega_{in} - \omega_0$, $\omega_{in}$ is the input laser frequency, $\omega_0$ is the cavity resonance frequency, $\kappa$ and $\kappa_{ext}$ are the total loss rate of the WGM and the coupling rate between the optical mode and the tapered fiber. The input power launched into the cavity is:

$$P = \hbar\omega_0 \left\langle \hat{a}_{in}^{\dagger}\hat{a}_{in} \right\rangle \tag{2.42}$$

In the classic limit for sufficiently large photon, the operator $\hat{a}$ can be replaced by the normalized complex light amplitude $\alpha$, and the dynamic equation can be written as:

$$\dot{\alpha} = i\Delta\alpha - \frac{\kappa}{2}\alpha + \sqrt{\kappa_{ext}}A_{in} \tag{2.43}$$

where $A_{in}$ represents the input field.

If the amplitude of input field is independent of time, by setting the time derivative to zero, the steady state solution of the intracavity field can be obtained as:

$$\alpha = \frac{\sqrt{\kappa_{ext}}A_{in}}{-i\Delta + \kappa/2} \tag{2.44}$$

In steady state, the power stored in the cavity is:

$$|A|^2 = \frac{|\alpha|^2}{\tau_{rt}} = \frac{2\kappa_{ext}}{\kappa}\frac{\mathcal{F}}{\pi}\frac{1}{1 + 4\Delta^2/\kappa^2}|A_{in}|^2 \tag{2.45}$$

where $\tau_{rt}$ is the round-trip time of photon in the cavity, and $\mathcal{F}$ is the fineness of the microcavity, which is defined as:

$$\mathcal{F} = \frac{\text{Spectral free path}}{\text{Optical mode linewidth}} = \frac{c}{nR\kappa} \tag{2.46}$$

The fineness is an important parameter to characterize the performance of micro-cavity, which can be seen from the Eq. (2.45) that when $\kappa = 2\kappa_0$ and $\Delta = 0$, the enhancement of intracavity power compared to launched power is $|A/A_{in}|^2 = \mathcal{F}/\pi$.

The transmitted light field, that is, the total output light field after coupling with the microcavity is:

$$A_{out} = A_{in} - \sqrt{\kappa_{ext}}\alpha \tag{2.47}$$

$$= \frac{2i\Delta + \kappa_0 - \kappa_{ext}}{2i\Delta + \kappa_0 + \kappa_{ext}}A_{in} \tag{2.48}$$

Therefore, the transmitted power at steady state can be obtained:

$$T = \left|\frac{A_{out}}{A_{in}}\right|^2 = \left|\frac{\kappa_0 - \kappa_{ext}}{\kappa_0 + \kappa_{ext}}\right|^2 \tag{2.49}$$

For a silica microsphere coupled to tapered fiber, their coupling strength $\kappa_{ext}$ can be continuously adjusted by tuning the taper-resonator gap, which is usually divided into three region based on the difference in coupling strength.

Undercoupling

The intrinsic damping rate of the microcavity is dominant in the total damping rate, $\kappa_0 > \kappa_{ext}$. The field coupling into the microcavity is smaller than that passing through the microcavity.

Critical Coupling

The intrinsic damping rate of the microcavity is equal to coupling rate to the tapered fiber. At the same time, the field emitted from the optical mode has the same amplitude as the field passing through the microcavity, with a phase difference of $\pi$, resulting in zero transmitted power. Critical coupling is analogized to impedance matching in electronic circuits, describing the external RF signal is completely coupled into the device without reflection.

Overcoupling

The intrinsic damping rate of the WGM is smaller than the coupling associated with the waveguide-resonator interface, $\kappa_0 < \kappa_{ext}$.

## References


1. Strutt J (1910) Philos Mag 20:1001
2. Schliesser A (2009) Cavity optomechanics and optical frequency comb generation with silica whispering-gallery-mode microresonators. PhD thesis
3. Park YS (2009) Radiation pressure cooling of a silica optomechanical resonator. PhD thesis
4. Dong CH (2011) Experiment of whispering gallery mode microcavity and cavity quantum dynamics. PhD thesis
5. Zou CL (2013) Novel integrated photonic devices for quantum information processor. PhD thesis
6. Borselli M, Johnson TJ, Painter O (2005) Opt Express 13(5):1515
7. Gorodetsky ML, Savchenkov AA, Ilchenko VS (1996) Opt Lett 21(7):453
8. Gorodetsky ML, Pryamikov AD, Ilchenko VS (2000) J Opt Soc Am B - Opt Phys 17(6):1051
9. Min B, Yang L, Vahala K (2007) Phys Rev A 76(1):013823
10. Vernooy DW, Ilchenko VS, Mabuchi H, Sreed WW, Kimble HJ (1998) Opt Lett 23(4):247
11. Rokhsari H, Spillane SM, Vahala KJ (2004) Appl Phys Lett 85(15):3029
12. Cleland ANN (2003) Foundations of nanomechanics. Springer, Berlin
13. Ekinci KL, Roukes ML (2005) Rev Sci Instrum 76(6):061101
14. Aspelmeyer M, Kippenberg TJ, Marquardt F (2014) Rev Mod Phys 86(4):1391
15. Little BE, Laine JP, Haus HA (1999) J Lightwave Technol 17(4):704

# Chapter 3
# Optomechanical Interaction

**Abstract** Based on the experiment of optomechanically induced transparency and light storage, we describe the optomechanical interaction in microcavity and related measurement techniques, such as the Pound–Drever–Hall (PDH) frequency stabilization technique, heterodyne detection and the gated-detection of mechanical vibrations, which are crucial for the subsequent experiments. The theoretical model of photon-phonon interaction in the optomechanical system is discussed in details. Moreover, we experimentally study the shifted frequency of cavity mode induced by Kerr effect with the driving power. And we can compensate the Kerr effect by appropriately shifting the locking frequency.

Optomechanics is an interesting field exploring the interaction between light and mechanical vibration. The optical forces are typically weak on macroscopic mechanical mirror. Therefore, the optical cavities are introduced to enhance radiation pressure force [1, 2]. In such an optomechanical resonator, optomechanically induced transparency (OMIT) can be used to characterize the effective optomechanical coupling of optical and mechanical excitation, which is the optomechanical analogy of the well-known phenomenon of electromagnetically induced transparency (EIT) [3]. The OMIT has been demonstrated in a variety of optomechanical systems, such as whispering gallery modes microcavity [4–7], photonic crystal nanocavity [8–10], membrane [11], and nanoelectromechanics [12].

During the OMIT process, a coherent mechanical oscillation is excited through optomechanical coupling by a strong control field at red sideband coupling, which is one mechanical frequency below the optical cavity resonance. The red sideband coupling leads to interconversion between photons and phonons. Anti-Stokes scattering of the control field off this mechanical excitation then generates an intracavity optical field at the cavity resonance. OMIT arises from the destructive interference between the probe field and the anti-Stokes scattering of the control field from this mechanical excitation. The interactions during the OMIT processes play an important role in optomechanical systems such as slow light [8], optomechanical light storage [8, 13, 14], optical wavelength conversion [5, 6, 15–17].



Z. Shen, *Experimental Research of Cavity Optomechanics*, Springer Theses,
https://doi.org/10.1007/978-981-33-4458-7_3

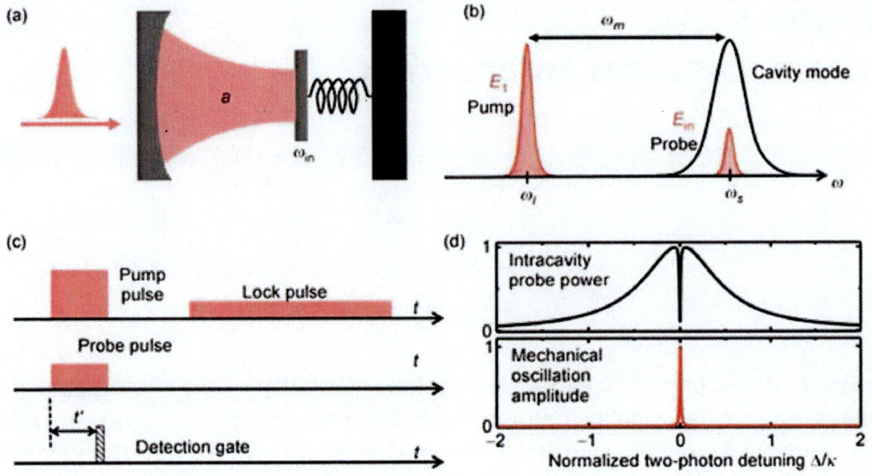


**Fig. 3.1** **a** Schematic of an optomechanical resonator. **b** Spectral position for the driving and signal pulses used in optomechanical induced transparency experiment. The strong control laser on the red sideband drives the optomechanical coupling, while a weak probe laser scans across the cavity resonance. **c** Schematic of the pulse sequence and the detection gate used for the experiments. **d** The theoretically expected intracavity probe power, oscillation amplitude in the steady state

## 3.1 Theoretical Analysis of Optomechanical Interaction

First the mechanism of photon-phonon conversion in the optomechanical system is studied theoretically. As shown in Fig. 3.1a, a mechanical oscillator is coupling to an optical cavity mode, with the cavity driven by a strong external laser field at the red sideband. The control light frequency $\omega_l$ is one mechanical frequency $\omega_m$ below the cavity resonance $\omega_c$, as shown schematically in Fig. 3.1b. In the resolved-sideband limit ($\omega_m > k$, where $k$ is the optical mode damping rate), the linearized optomechanical Hamiltonian can be described by:

$$H = \hbar\omega_c \hat{a}^\dagger \hat{a} + \hbar\omega_m \hat{b}^\dagger \hat{b} + \hbar G(\hat{a}^\dagger \hat{b} + \hat{a}\hat{b}^\dagger), \tag{3.1}$$

where $\hat{a}$ and $\hat{b}$ are the annihilation operator for the intracavity signal field and the mechanical mode, respectively, and $G$ is the effective optomechanical coupling rate determined by the control laser. As shown by this Hamiltonian, the cavity field couples the mechanical displacement, which can induce a coherent interconversion between the optical and motional states. Therefore, this system can be used for optomechanical quantum state transfer and frequency conversion [18, 19].

With optical and mechanical damping included, the optomechanical coupling between the signal field and mechanical displacement can be described by the following equations of motion:

$$\dot{\alpha} = [i(\omega_{in} - \omega_0) - k/2]\alpha - iG\beta + \sqrt{k_{ext}}A_{in}, \tag{3.2}$$

$$\dot{\beta} = [i(\omega_{in} - \omega_0) - \gamma_m/2]\beta - iG\alpha. \tag{3.3}$$

where $\omega_{in}$ and $A_{in}$ are the frequency and amplitude of the input optical probe field, respectively, and $k$ and $k_{ext}$ are the total cavity decay rate and the effective output coupling rate of the probe field, respectively, $\gamma_m$ is the mechanical damping rate. Note that $\alpha$ and $\beta$ are normalized such that $|\alpha|^2$ is intracavity photon number and $|\beta|^2$ is the phonon number. The coherent mechanical excitation is generated through the interconversion process mentioned above. Destructive interference between the intracavity fields generated by the anti-Stokes scattering and that generated by the input signal prevents the excitation of the optical mode, leading to a transparent dip (OMIT) in the optical spectra. As will be discussed in detail later, both the OMIT and the optomechanical light storage can be described by the coupled oscillator equations. In the steady state, the solution of equations are then given by:

$$\alpha = \frac{(i\Delta - \gamma_m/2)}{(i\Delta - k/2)(i\Delta - \gamma_m/2) + G^2}\sqrt{k_{ext}}A_{in}, \tag{3.4}$$

$$\beta = \frac{iG}{(i\Delta - k/2)(i\Delta - \gamma_m/2) + G^2}\sqrt{k_{ext}}A_{in}. \tag{3.5}$$

where $\Delta = \omega_{in} - \omega_0$ is the detuning between the external driving field and the optical cavity resonance. A typical power spectrum of intracavity probe power $|\alpha|^2$ and oscillation amplitude $|\beta|^2$ are shown in Fig. 3.1d. In the limit that $\gamma_m << k$, the bandwidth of the transparency window is determined by $\gamma_m + 4G^2/k = (1 + C)\gamma_m$. Here, $C = \frac{4G^2}{k\gamma_m}$ is the optomechanical cooperativity, a dimensionless parameter that characterizes the strength of optomechanical coupling. The relative depth of the OMIT dip is given by $\frac{1}{(1+C)^2}$.

## 3.2 Experimental Method

Experimental studies were carried out at room temperature. The silica microsphere was fabricated by melting the tapered tip of a fiber using a CW $CO_2$ laser. The coupling strength between the tapered fiber and the microsphere can be adjusted by changing the air gap between them, which was controlled by a high resolution and stable translation stage, as shown in Fig. 3.2a. The coupling system was held in a cleaning chamber in order to avoid contamination. The half wave plate was used to change the polarization of input laser. To observe the mechanical vibrations through tapered-fiber excitation, a CW input power far below the threshold for parametric instability is used.

Figure 3.2b shows the mechanical vibration power spectrum of a microcavity with a diameter of 35 μm, which is taken by an electronic spectrum analyzer (FSV5). There are several narrow peaks are identified as the radial breathing modes with (l, m) =

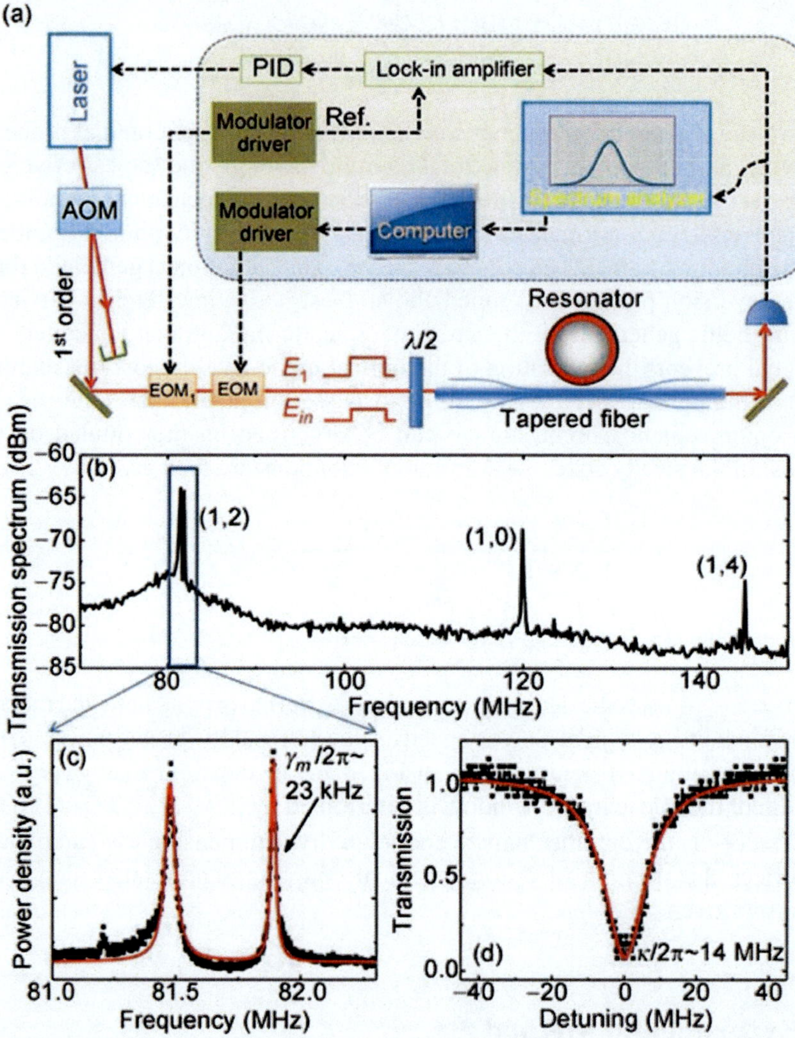


**Fig. 3.2** **a** Schematic of the experimental setup for the transient OMIT experiment, with solid red lines representing the optical paths and dashed black lines representing the electrical connections. The silica microresonator and the tapered fibre were held in a chamber in order to avoid contamination. AOM, EOM, λ/2 are acousto-optic modulator, electro-optic modulator, and half wave plate, respectively. **b** Displacement power spectrum of the three lowest radial breathing mode obtained from the microsphere with diameter of 35 μm, including three radial mechanical vibration modes. **c** Mechanical mode splitting was shown with an expanded frequency scale. **d** Transmission resonance for the WGM at 780 nm with $\kappa/2\pi = 14$ MHz

(1, 2), (1, 0) and (1, 4), respectively [20]. These modes can be probed by optical field because their vibration deforms the microsphere and modulate cavity path length in the equatorial plane. Then the change of the resonance frequency of the optical mode caused by vibration will modulate the transmittance and phase of the light field, and finally be observed by the spectrum analyzer. Due to the slight asymmetric shape of the microsphere, a mechanical mode splitting of 0.43 MHz was observed for (1, 2) mode, as shown in the details of Fig. 3.2. We choose the higher $Q$ mode that frequency is $\omega_m/2\pi = 81.9$ MHz and linewidth is $\gamma_m/2\pi = 23$ kHz for photon-phonon conversion experiments. We couple this mechanical mode to an optical whispering gallery mode (WGM) near $\lambda \approx 780$ nm and with a linewidth of $k/2\pi = 14$ MHz, as illustrated in Fig. 3.2d. With higher input power, the asymmetric lineshape of the optical mode can be observed [21, 22]. Thus, pulses sequence are used in the OMIT experiment to avoid the thermal bistability in such silica microresonator.

As illustrated in Fig. 3.2a, the control pulse and locking pulse were derived from the first-order diffraction from an acoustic-optic modulator (AOM). The control pulse also propagated through an electro-optic modulator (EOM). The blue sideband generated from the EOM with modulation frequency $\Omega$ served as the probe pulse. The frequency of the tunable laser was locked to the relative red sideband of the WGM in the microresonator by the Pound–Drever–Hall (PDH) frequency stabilization technique, where the locking pulse propagated through the $EOM_1$. To measure the optical emission from the microresonator near the WGM resonance, heterodyne detection was used with the control pulse serving as the local oscillator. To probe the mechanical mode, we have used a technique based on the gated-detection of mechanical vibrations, for which we use a weak probe pulse to measure the mechanical displacement power spectrum within a given time duration.

Figure 3.3a shows the pulse sequence from the detector without the EOM modulation. The first great pulse is only control pulse with frequency $\omega_l$ to drive the mechanical coupling. The second long pulse is the lock pulse which is synchronous to the $EOM_1$. When $EOM_1$ is on, we can lock the laser to the relevant optical mode. The measured Pound–Drever error signal is shown in the inset of Fig. 3.3. Turning on the EOM which is synchronous to the first pulse, we can run the OMIT experiment.

First, the photon-phonon interconversion process is studied by measure the heterodyne beating between the control pulse and the probe pulse directly, with a digital oscilloscope. The time domain signal directly reflects the process of photon-phonon conversion. Figure 3.3b shows the heterodyne beats as a function of time, obtained with incident powers for the control laser is $P_c = 3$ mW, corresponding to $G/2\pi \approx 0.32$ MHz. The duration of the controlling pulse is 9 $\mu$s. The relatively long controlling pulse allowed us to probe the temporal behavior of the interconversion between the optical and the mechanical excitations. The power of signal field is about $P_{sig} = 0.01$ mW. The inset of Fig. 3.3b plots a portion of the beats with an expanded time scale and also compares the experimental result with a oscillation period consistent with the mechanical frequency. The amplitude of the periodic oscillation is decreased with the time, approaching the steady state of the coupled optomechanical system.

In addition, the transient behavior of the optical emission was measured with a spectrum analyzer operating in a gated-detection mode with different resolution bandwidth (RBW). Figure 3.3c plots the optical emission power from the WGM as a function of the detection time at the modulation frequency $\Omega = \omega_m$. The temporal behavior of the optical emission is similar to the evolution of oscillation amplitude

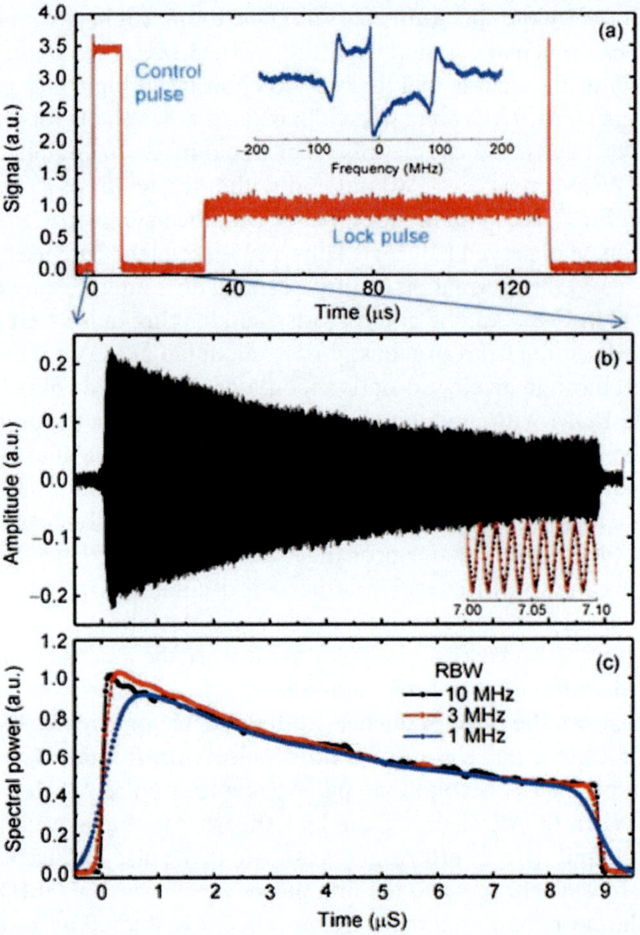


**Fig. 3.3** **a** The pulse signal measured directly from the detector. Inset: measured Pound–Drever–Hall error signal. **b–c** The heterodyne beating between the control pulse and the probe pulse as a function of the timing with $\Omega = \omega_m$. **b** is measured by an oscilloscope, **c** shows the beats obtained from the digital oscilloscope and **c** shows the signal obtained from the electronic spectrum analyzer

in Fig. 3.3b. The resolution bandwidth was set to 1, 3 or 10 MHz. The higher RBW shows the sharp edge electric signal.

Figure 3.4 shows the typical emission power as a function of the modulation frequency $\Omega$. The dip in the emission power observed near $\Omega = \omega_m$ indicates the OMIT. For comparison with theoretical calculations, optical emission powers are normalized to that obtained in the absence of the control and with the signal pulse at the WGM resonance. The resolution bandwidth of the gated-detection is 3 MHz with gate delay time $t' = 8.5$ μs and gate length 0.5 μs. The OMIT dip is coincident with the emission power at the end of the probe pulse in Fig. 3.3b.

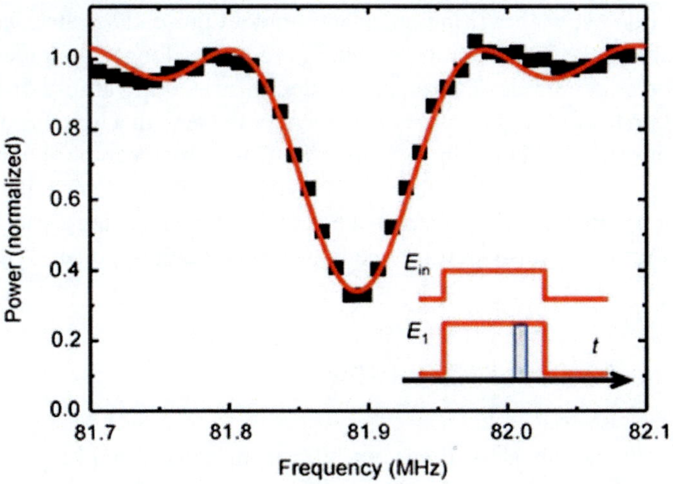


**Fig. 3.4** The typical optomechanically induced transparency observed in the experiment. The result was obtained as a function of the detuning between the signal field and the control field, which is fixed at the red sideband. Solid curves are the theoretically calculated spectral response. The mechanical coupling rate of the control pulse is estimated as $G/2\pi \approx 0.32$ MHz. The inset shows the pulsed pump and gate detection scheme in the experiment

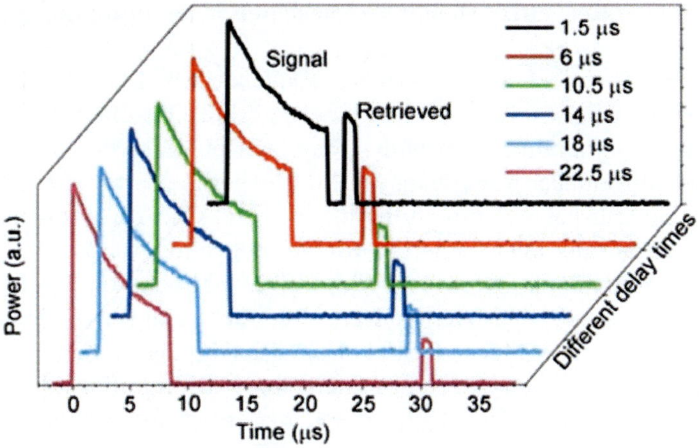


**Fig. 3.5** **a** The power of the signal and retrieved pulse emitted from the microsphere and measured with time-gated heterodyne detection and with different delay time between the writing and readout pulse. From top to down, the delay time are 1.5, 6, 10.5, 14, 18, 22.5 μ

To clearly understand the intraconversion between the optical motion and mechanical motion, another laser pulse following the control pulse is induced to convert the mechanical motion to the optical motion [13, 14]. This process has also been called signal readout for the light storage experiment. Two separated control pulses with durations of 9 and 1 μs served as the writing and readout pulse for the experiment. Figure 3.5 shows the experimental results of normalized heterodyne traces with different delay time between writing pulse and readout pulse. The retrieved signal can even be observed after 22 μs storage, which are orders longer than the life time of photons in the cavity mode. Therefore, we demonstrated the ability of the optomechanical system to store light pulse and to recover these with a long decay time up to 22 μs of the mechanical coherence. Longer optical storage can be achieved through narrower linewidth mechanical vibration modes, such as nanomechanical oscillator [23]. In such optomechanical system, the Ramsey interferometry can be achieved when one more probe pulse is introduced during the reading pulse [24].

## 3.3 Compensation of Kerr Effect

For the study of the mechanical system, the strong optical field is usually used to drive the mechanical motion at the sideband of the cavity modes. Therefore, the Kerr effect and thermal effect, induced by the intracavity field, leads to the frequency shift of the optical mode, which also can be used for the all-optical switching [25–27]. If the driving field is continuous, the frequency shift should be constant during the experiment. However, the thermal effect gives rise to the unstable in the system. To weaken the thermal instability, the pulsed laser is introduced to drive the mechanical motion, while the Kerr effect should be considered in an ultra-high Q microcavity [25].

In the high-Q microcavities the cavity frequency was red shifted by the Kerr effect by the laser pulse even with low power [25]. This shifted frequency would exceed over tens to hundred MHz with strong laser pulse, as shown in Fig. 3.6a. In transient optically induced transparency (OMIT) experiments, the Kerr effect induces the asymmetric transparency dip during the transient OMIT experiment. However, the symmetric transmission is important for the application based on the mechanical system, such as circulator and isolator [28, 29]. In addition, to a mechanical mode with frequency around 100 MHz, the Kerr effect of strong driving pulse may lead the cavity mode from the blue sideband of the driving pulse to the red sideband. In this case, the effects of optomechanical interactions will significantly change.

Here, we present the Kerr effect induced by strong driving pulse in the silica microspheres. We experimentally demonstrate the shifted frequency of cavity mode induced by Kerr effect scales linearly with the driving power. The rate is −0.5 MHz/mW. Minus represents red-shift of the cavity resonance frequency. The Kerr effect leads to the asymmetric transparency dip during the transient OMIT experiment. Under the known driving power, we could change the locking frequency depending on the rate to compensate the Kerr effect and show the symmetric transparency window again. It is important for the study of the inter-conversion between photon and phonon especially with strong driving power [6, 9, 30, 31].

The strong optical field is usually used to enhance the optomechanical coupling at the red sideband of the cavity modes. To eliminate the heating induced by the strong driving fields, the duty cycle of the experiment is kept 6% by using 12 μs optical pulses. For the transient OMIT experiment, both the driving and the signal pulses

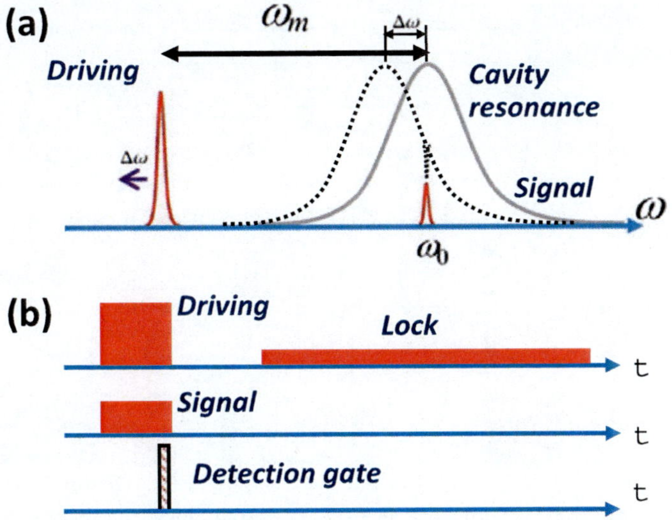


**Fig. 3.6** **a** The influence of the Kerr effect for transient OMIT. The dashed black line shows under the driving pulse the intracavity signal power spectra is asymmetric due to the Kerr effect, which can be compensated by shifting the driving frequency. **b** Schematic of the pulse sequence and the detection gate used for the transient OMIT experiments

came from a single-frequency tunable diode laser and with its frequency locked to the WGM resonance using the Pound–Drever–Hall technique with the weak locking pulse. The driving pulse was fixed at $\omega_m$ below a given WGM resonance and the incident signal pulse was tuned to near the WGM resonance as shown in Fig. 3.6a. The dashed black line represents under the driving pulse the intracavity signal power spectra is asymmetric, which is because of the WGM resonance frequency red-shift, $\Delta\omega$, induced by the Kerr effect. At this time, we can reproduce the symmetrical intracavity power spectrum and compensate the Kerr effect by shifting the driving frequency of $\Delta\omega$.

As illustrated in the schematic of the experimental setup in Fig. 3.7, the strong driving pulses and the weak locking pulse were derived from the first-order diffraction from an acoustic-optic modulator ($AOM_1$). The driving pulse propagated through an electro-optic modulator ($EOM_1$) with the modulation frequency around $\omega_m$. The blue sideband generated from the $EOM_1$ served as the signal pulse. Both the driving and the signal pulses were square shaped, with the same timing and with a duration of 12 $\mu$s. Heterodyne detection was used for the measurement of the optical emission from the microsphere near the WGM resonance, with the driving pulse serving as the local oscillator. Since the incident signal pulse is generated directly from the local oscillator with a pure phase modulation, the heterodyne detection is not sensitive to the transmitted signal pulse that is not emitted from the optical resonator. In this regard, the heterodyne detection directly measures the power of the intracavity optical field. Similar techniques were also used in earlier OMIT and optomechanical light storage studies. The transient behavior of the optical emission was probed

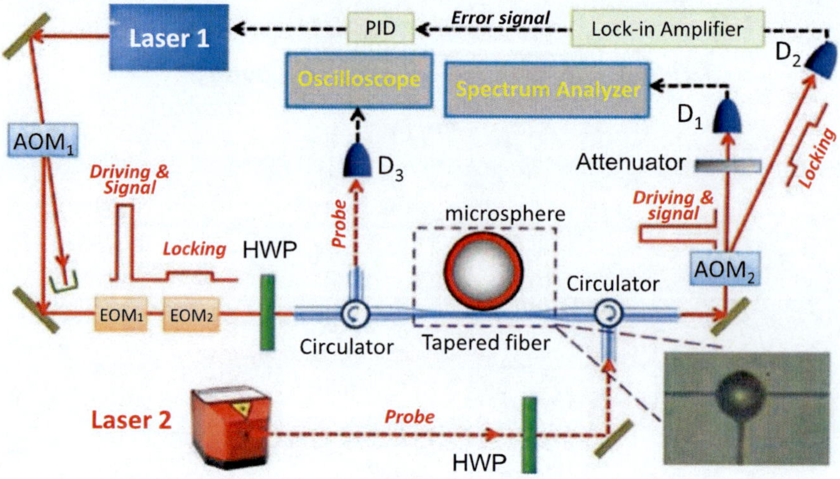


**Fig. 3.7** **a** Schematic of the experimental setup. The red solid lines represent the optical paths of driving, signal and locking pulse. The red dashed lines represent the probe laser coupled through the double circulators from the opposite direction and dashed black lines represent the electrical connections. The silica microresonator and the tapered fibre were held in a cleaning chamber in order to avoid contamination. AOM, EOM, D, and HWP are acousto-optic modulator, electro-optic modulator, detector, and half wave plate, respectively

with a spectrum analyzer operating in a time-gated detection mode, with the time resolution set by the resolution bandwidth of 3 MHz. To avoid the saturation of the photo detector simultaneously exposed to the strong driving pulse and the weak locking pulse, $AOM_2$ was introduced to separate these two pulses. The error signal could be obtained for the PID controller for locking the laser at $f$ below a given WGM resonance when the locking pulse passed through $EOM_2$ with the modulation frequency $f$.

First, to study the temporal Kerr effect, we input another probe laser (Laser 2) which was resonant with the given WGM through the opposite direction using the double circulators, as shown in Fig. 3.7. In our experiment, the optical field of a WGM was excited through the evanescent field of a tapered optical fiber. The silica microspheres with diameter $D$ of 38 $\mu$m were fabricated by melting the tapered tip of a fiber through a continuous wave $CO_2$ laser. The tapered fiber was obtained by melting a standard SMF-28 silica optical fiber with a hydrogen flame. Both the silica microsphere and the tapered fiber were held in an enclosed chamber in order to avoid spoiling of optical $Q$ factors.

Figure 3.8a shows the typical transmission of the WGM with linewidth of 22 MHz, corresponding to $Q \sim 1.7 \times 10^7$. Figure 3.8b shows the frequency drifting during the driving power pulse. The black line is transmission when the Laser 2 frequency is resonance to the WGM and Laser 1 is lock at the red sideband of a given WGM resonance. It is clear that the transmission is changed during the strong driving pulse with 6 mW. As sketched in Fig. 3.8a, the shifted frequency can be estimated from the transmission and the linewidth. The ratio of the frequency shift

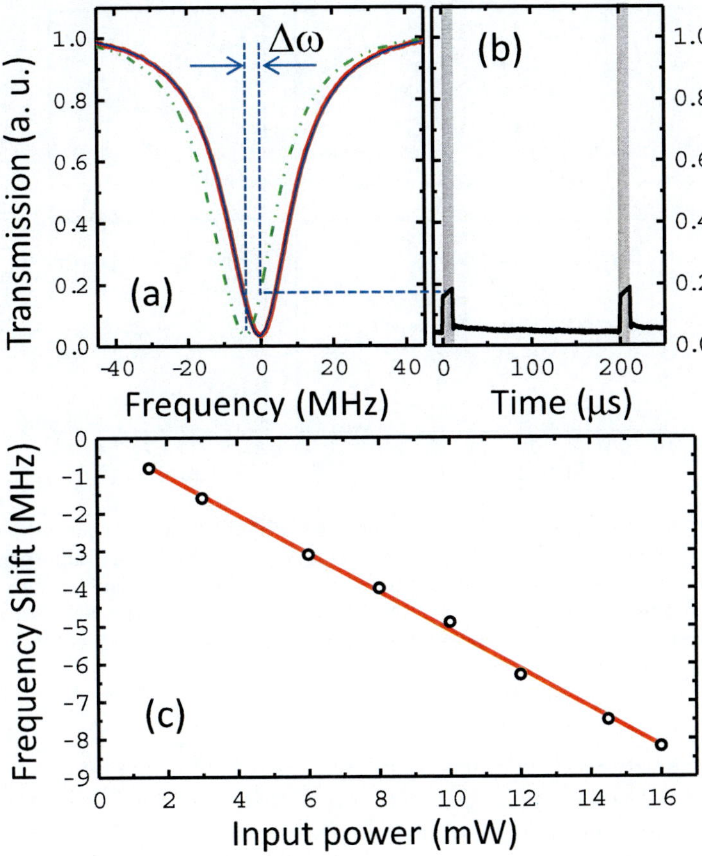


**Fig. 3.8** **a** The typical transmission resonance for the WGM at 780 nm, with $\kappa/2\pi = 22$ MHz, corresponding to $Q \sim 1.7 \times 10^7$. The green dash line show temporal Kerr shift during the driving pulse. **b** The black line represents the signal output from detector $D_3$. The gray area indicates that the strong driving light is launched. **c** Frequency drift versus input driving power. The red line is fitting results with the slope of −0.5 MHz/mW

of the Kerr effect versus the input power is –0.5 MHz/mW, as shown in Fig. 3.8c. This result is coincident to the calculation of Kerr shift induced by the intracavity photon number $\Delta\omega = -\omega_c \frac{n_2 P_c}{n_0 S}$, where $n_0$ and $n_2$ are the refractive index of silica and the optical Kerr coefficient, respectively, and $S$ is the effective cross-sectional area of the relevant WGM. $P_c$ is the intracavity optical power which can be given as $P_c = \frac{\lambda_0}{\pi^2 n_0 D} \frac{1}{1+(2\Delta/\kappa)^2} \frac{Q^2}{Q_{ex}} P_{in}$ [32, 33], where $\lambda_0$ and $\Delta$ are the resonance wavelength and the detuning between the driving field and the optical mode. Finally, $Q_{ex}$ is the external quality factor determined by the coupling between the resonator and the tapered fiber.

Figure 3.9 shows the emission power from the microcavity as a function of the detuning, $\delta$, between the signal pulse and the driving pulse. The transparency window in the emission power observed near $\delta = \omega_m$ indicates the underlying OMIT process.

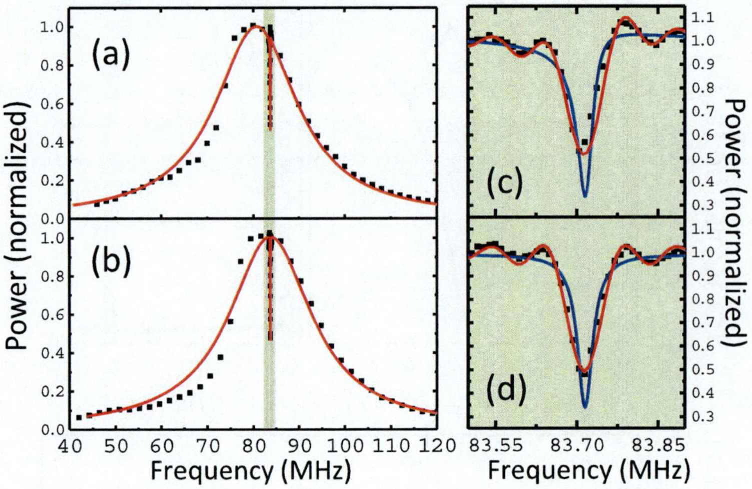


**Fig. 3.9** **a–b** The emission power spectra are obtained with the locking frequency $f$ of 83.7 and 86.7 MHz, respectively. The incident driving power is 6 mW. **c–d** The spectral response near $\omega_m$ with an expanded frequency scale. The red lines are the results of calculations using the parameters mechanical frequency $\omega_m/2\pi = 83.7$ MHz, the optical mode linewidth $\kappa/2\pi = 22$ MHz, mechanical mode linewidth $\gamma_m/2\pi = 20$ kHz, optomechanical coupling rate $G/2\pi = 0.28$ MHz, and $\Delta\omega/2\pi = 3$ MHz respectively

The difference between OMIT spectra in Fig. 3.9a is the locking frequency $f$, which are 83.7 and 86.7 MHz, respectively. The asymmetric spectra in Fig. 3.9a is due to the WGM resonance frequency red-shift $\Delta\omega$. Therefore, the transparency is appeared symmetric when the locking frequency shift to 86.7 MHz, which is 3 MHz larger than mechanical frequency. And the detuning $\Delta\omega$ is coincident with the input power of 6 mW. Of course, the signal pulse, which is resonant to the cavity mode, is low enough to avoid the Kerr effect for our measurement. The expanded curves in Fig. 3.9c, d clearly show the asymmetric dip and symmetric dip in the corresponding OMIT spectra. It also demonstrated that the frequency of given WGM is shifted during the driving pulse. The results are in good agreement with the theoretically calculated spectrum of the intracavity power as a function of the detuning between the signal and the driving, as shown by the solid red lines in Fig. 3.9 [34]. The reason of oscillations in Fig. 3.9c, d is that the intracavity signal field is unable to reach steady state with 12 $\mu$s driving light. The solid blue lines in the Fig. 3.9c, d represent the steady state of the optomechanical interaction.

With low driving power, it's hard to observe the slight asymmetric shape of the transparency dip. For stronger driving power, the drift of WGM is greater and the asymmetry of the OMIT dip is increased, as shown the Fano shape in Fig. 3.10a. Here, the input power is 16 mW. To observe the symmetric OMIT dip, we should shift the locking frequency from 83.7 to 92.7 MHz. As shown in Fig. 3.10b, the symmetric OMIT dip is obtained again. It's note that the shifted frequency of 9 MHz is more than the estimated value of 16 mW. We believe that it derives from the contribution of

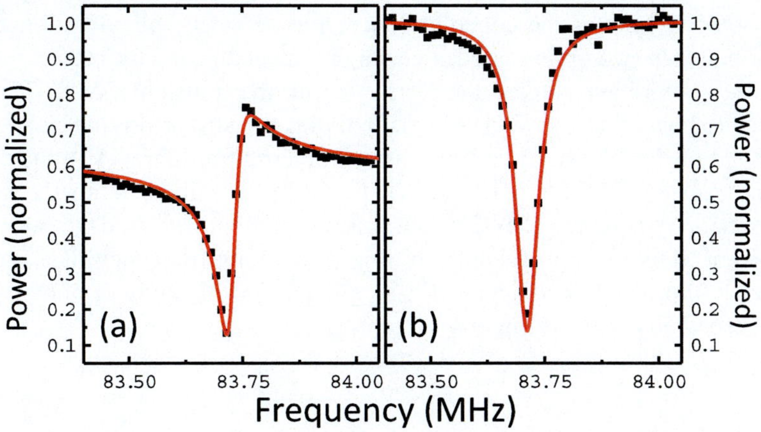


**Fig. 3.10** The asymmetric and symmetric OMIT dip are obtained before and after the shifted locking frequency of $\omega_m$. Here, the incident driving power used is 16 mW. The red lines are the results of calculations using the parameters $\kappa/2\pi = 22$ MHz, $\gamma_m/2\pi = 20$ kHz, $G/2\pi = 0.46$ MHz, $\Delta\omega/2\pi = 9$ MHz, respectively

thermal effect, which is also shown as the slow drift of frequency during the driving pulse in Fig. 3.8b.

## 3.4 Summary

Cavity optomechanics provides a unique platform for the study of the interconversion between optical and mechanical excitations. Due to the thermal excitations of the mechanical oscillator, there are noises affecting the coherent interconversion. For example, the thermal mean phonon number of our experiment is determined by $n_{th} \approx k_B T/\hbar\omega_m \approx 7.6 \times 10^4$. For the classical applications that cavity mean phonon number is much larger than $n_{th}$, the thermal mechanical noise can be negligible. For the application of the interconversion process in quantum regime [35, 36], the mechanical oscillator should be cooled to its quantum ground state, in which the thermal mechanical motion can be eliminated [20, 30, 31, 37–39]. Effects of thermal mechanical motion can also be circumvented with the use of an optomechanical dark mode, as realized by coupling two, instead of one, optical modes to a mechanical oscillator [5, 40, 41].

In summary, experimental studies of OMIT responses probed with a gated-trigger detection technique provide detailed information to help us understanding the transient behavior of OMIT in the silica microresonator. These studies confirm the coherence interconversion between the photons and phonons. The characteristic time scale for establishing transparency window is $[(1 + C)\gamma_m]^{-1}$. In transient optomechanical processes, this time scale should also be the characteristic time scale for the application of OMIT, such as the coherent conversion between optical and mechanical modes or between two optical modes via an optomechanical dark mode.

In addition, we have studied the Kerr effect in the transient OMIT. The experiment result shows that the WGM is drifted with the strong driving pulse even at the sideband of the WGM. Therefore, the asymmetric transparency dip is observed during the transient OMIT experiment. We pay close attention to study the Kerr shift of the resonance frequency of WGM with various input power. And we can compensate the Kerr effect by appropriately shifting the locking frequency. The symmetric transparency dip is obtained again. It is significant for the study of photon-phonon coupling especially with strong driving power.

## References

1. Kippenberg TJ, Rokhsari H, Carmon T, Scherer A, Vahala KJ (2005) Phys Rev Lett 95(3):033901
2. Carmon T, Rokhsari H, Yang L, Kippenberg TJ, Vahala KJ (2005) Phys Rev Lett 94(22):223902
3. Agarwal GS, Huang SM (2010) Phys Rev A 81:041803
4. Weis S, Riviere R, Deleglise S, Gavartin E, Arcizet O, Schliesser A, Kippenberg TJ (2010) Science 330(6010):1520
5. Dong C, Fiore V, Kuzyk MC, Wang H (2012) Science 338(6114):1609
6. Liu Y, Davanço M, Aksyuk V, Srinivasan K (2013) Phys Rev Lett 110(22):223603
7. Fan L, Fong KY, Poot M, Tang HX (2015) Nat Commun 6:5850
8. Safavi-Naeini AH, Alegre TPM, Chan J, Eichenfield M, Winger M, Lin Q, Hill JT, Chang DE, Painter O (2011) Nature 472:69
9. Chan J, Alegre TPM, Safavi-Naeini AH, Hill JT, Krause A, Groeblacher S, Aspelmeyer M, Painter O (2011) Nature 478:89
10. Davanco M, Ates S, Liu Y, Srinivasan K (2014) Appl Phys Lett 104(4):041101
11. Karuza M, Biancofiore C, Bawaj M, Molinelli C, Galassi M, Natali R, Tombesi P, Giuseppe GD, Vitali D (2013) Phys Rev A 88(1):013804
12. Zhou X, Hocke F, Schliesser AMA, Huebl H, Gross R, Kippenberg TJ (2013) Nat Phys 9:179
13. Fiore V, Yang Y, Kuzyk MC, Barbour R, Tian L, Wang H (2011) Phys Rev Lett 107(13):133601
14. Fiore V, Dong CH, Kuzyk MC, Wang H (2013) Phys Rev A 87(2):023812
15. Hill JT, Safavi-Naeini AH, Chan J, Painter O (2012) Nat Commun 3:1196
16. Bochmann J, Vainsencher A, Awschalom DD, Cleland AN (2013) Nat Phys 9(11):712
17. Andrews RW, Peterson RW, Purly TP, Cicak K, Simmonds RW, Regal CA, Lehnert KW (2014) Nat Phys 10:321
18. Genes C, Mari A, Tombesi P, Vitali D (2008) Phys Rev A 78(3):032316
19. Tian L, Wang H (2010) Phys Rev A 82(5):053806
20. Park YS, Wang H (2009) Nat Phys 5:489
21. Carmon T, Yang L, Vahala KJ (2004) Opt Express 12(20):4742
22. He L, Xiao YF, Dong CH, Zhu J, Gaddam V, Yang L (2008) Appl Phys Lett 93(20):201102
23. Oo T, Dong C, Fiore V, Wang H (2013) Appl Phys Lett 103(3):031116
24. Qu K, Dong CH, Wang H, Agarwal GS (2014) Phys Rev A 90(5):053809
25. Yoshiki W, Tanabe T (2014) Opt Express 22(20):24332
26. Eckhouse V, Cestier I, Eisenstein G, Combrie S, Lehoucq G, De Rossi A (2012) Opt Express 20(8):8524
27. Pelc JS, Rivoire K, Vo S, Santori C, Fattal DA, Beausoleil RG (2014) Opt Express 22(4):3797
28. Hafezi M, Rabl P (2012) Opt Express 20(7):7672
29. Fu W, Shu FJ, Zhang YL, Dong CH, Zou CL, Guo GC (2015) Opt Express 23(19):25118

30. Verhagen E, Deleglise S, Weis S, Schliesser A, Kippenberg TJ (2012) Nature 482:63
31. Liu YC, Xiao YF, Luan X, Wong CW (2013) Phys Rev Lett 110(15):153606
32. Park YS, Wang H (2007) Opt Lett 32(21):3104
33. Xiao YF, Ozdemir SK, Gaddam V, Dong CH, Imoto N, Yang L (2008) Opt Express 16(26):21462
34. Dong C, Fiore V, Kuzyk MC, Wang H (2013) Phys Rev A 87(5):055802
35. Tian L (2012) Phys Rev Lett 108(15):153604
36. Liu YC, Xiao YF, Chen YL, Yu XC, Gong Q (2013) Phys Rev Lett 111(8):083601
37. Teufel J, Donner T, Li D, Harlow J, Allman M, Cicak K, Sirois A, Whittaker JD, Lehnert K, Simmonds RW (2011) Nature 471:204
38. Palomaki T, Harlow J, Teufel J, Simmonds R, Lehnert K (2013) Nature 495(7440):210
39. Galland C, Sangouard N, Piro N, Gisin N, Kippenberg TJ (2014) Phys Rev Lett 112(14):143602
40. Wang YD, Clerk AA (2012) New J Phys 14(10):105010
41. Wang YD, Clerk AA (2012) Phys Rev Lett 108(15):153603

# Chapter 4
# Experimental Demonstration of Optomechanically Induced Non-reciprocity

**Abstract** Non-reciprocal devices, such as circulators and isolators, which allow light to pass in specified direction but block light in the undesired direction, are indispensable components in classical and quantum information processing in photonic integrated circuits. However, the integration of the conventional optical non-reciprocal devices, which employ the magneto-optical Faraday effect, is technically challenging because of material incompatibilities and the large footprint of such device designs. In recent years, there has been growing interest in the realization of non-reciprocal photonic devices without magnetic materials. Here, we experimentally demonstrated non-magnetic non-reciprocity using optomechanical interactions in a whispering-gallery microresonator, as proposed by Hafezi and Rabl. In this approach, the travelling wave nature of the microresonator produces a rise in the degenerate clockwise (CW) and counter-clockwise (CCW) travelling-wave optical modes. For the optomechanical interaction, the CW and CCW modes are independently coupled with the mechanical mode. Only when the optical driving and signal fields are coupled to the same optical mode, the coherent conversion between signal photon and phonon is enabled. As a result, the directional driving field breaks the time-reversal symmetry and leads to non-reciprocal transmittance for the signal light. In a fiber coupled silica microsphere system, we observed optomechanically induced non-reciprocal transparency (OMIT) and amplification (OMIA), and demonstrated a non-reciprocal phase shift of up to 40°.

## 4.1 Introduction

Optical isolators allow light to pass in one direction but block light in the opposite direction. It can prevent reflected light from disturbing the operation of the laser or multipath interference in a complex optical communication system. The suppression of crosstalk between different devices and undesired light routing generally improved the system designability [1]. Therefore, optical isolators are indispensable components in classical and quantum information processing. The most common optical non-reciprocal devices are based on the Faraday effect of magneto-optical materials, and its working principle is shown in Fig. 4.1. For the forward incident



Z. Shen, *Experimental Research of Cavity Optomechanics*, Springer Theses,
https://doi.org/10.1007/978-981-33-4458-7_4

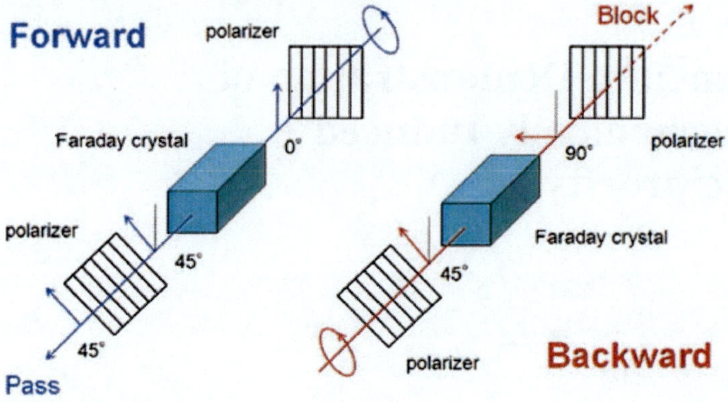


Fig. 4.1 Schematic of magneto-optical isolator

signal light, it becomes linearly polarized light after passing through the polarizer. The Faraday crystal biased by the external magnetic field rotates the polarization by 45° and pass another polarizer. For the backward signal light, The Faraday crystal rotates the polarization to horizontal, orthogonal to the polarizer. Thus the backward light is completely blocked.

Although magneto-optical materials can be used to realize optical isolators featuring excellent performance, they are difficult to integrate on-chip. The difficulties lie in the incompatibility between magneto-optical materials and traditional semiconductor materials, the requirement of external magnetic field bias, and the high transmission loss of magneto-optical materials [2]. Therefore, in recent years there has been growing interest in the realization of nonreciprocal photonic devices without magnetic materials. To create genuine non-reciprocal devices, it is necessary to break the Lorentz reciprocity [1]. Thus far, achieving this goal is primarily rely on the spatio-temporal modulations and the nonlinear optical effect [3–15].

We demonstrate the optomechanically induced non-reciprocity in a whispering gallery mode microsphere. This approach was first suggested by Hafezi and Rabl [12]. Due to the momentum conservation of phonon-photon interaction, only the photon propagating in certain direction can coherently couple with phonon, thus lead to the non-reciprocal response. In our experiment, we demonstrate the non-reciprocal optomechanically induced transparency (OMIT). Moreover, we report the experimental demonstration of optomechanically induced non-reciprocal phase shift for light passing through the optomechanical system. The non-reciprocity can also be realized through Brillouin scattering [7, 8], where two optical modes couples with a traveling acoustic mode. The triply resonant interact can enhance the interaction, while it is hard to fulfill the momentum and energy conservation conditions for the given mechanical mode in practical. In contrast, the underlying mechanism of optomechanically induced non-reciprocity can be generalized to any traveling optical wave resonators with mechanical vibration. It can be operated in a broaden wavelength range and readily to be implemented in current integrated photonic chips [16, 17]. With the mechanical vibrations being cooled to their ground states, applications in the quantum regime-such as single-photon isolators and circulators-also become possible.

## 4.2 Theoretical Derivation and Data Analysis

Figure 4.2 schematically illustrates a traveling wave optomechanical system that consists of a whispering gallery microresonator that is evanescently coupled with a tapered fiber. The traveling wave nature of the microresonator produces a rise in the degenerate clockwise (CW) and counter-clockwise (CCW) traveling-wave whispering-galley modes (WGMs), which can also be represented by the orbital angular momentums $m$ and $-m$, respectively. Similarly, the mechanical modes in such axial symmetric geometry have an orbital angular momentum $m_b$, where $m_b = 0$ represents the breath vibration mode, where the equator of the sphere expands and compresses uniformly. The variation of the cavity radius leads to the modification of the resonant frequency of all optical modes [18], which can be described by the dispersive optomechanical interaction Hamiltonian:

$$H_{int} = g_0(a_{\mathrm{cw}}^{\dagger}a_{\mathrm{cw}} + a_{\mathrm{ccw}}^{\dagger}a_{\mathrm{ccw}})(b + b^{\dagger}). \tag{4.1}$$

where $a_{\mathrm{cw(ccw)}}$ and $b$ denote the Bosonic operators of the CW(CCW) optical mode and mechanical mode, respectively; $g_0$ denotes the single photon coupling rate corresponding to the optical mode resonance shift caused by one phonon. For such an interaction, only a single optical mode (either $a_{\mathrm{cw}}$ or $a_{\mathrm{ccw}}$) is involved; therefore, the properties of the driving and signal optical fields should be identical, except for the frequency difference. Thus, the CW(CCW) driving field can only stimulate the interaction between a phonon and a CW(CCW) signal photon, as shown in

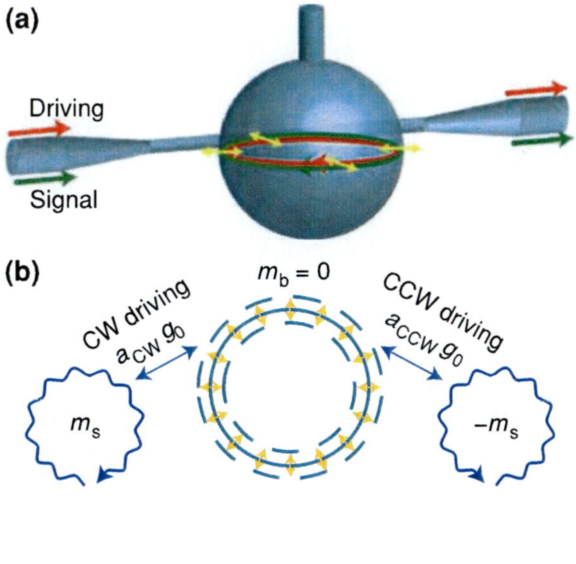


**Fig. 4.2** **a** A strong driving field enhances the optomechanical coupling between a vibrational mode and the co-propagating optical mode inside a microcavity. **b** The schematic of clockwise and counter-clockwise optical mode coupling with the breathing mode

Fig. 4.2b. This relation between the driving and signal directions can also be interpreted using the conservation of orbit angular momentum $m_s - m_d = m_b = 0$, where $m_s$ and $m_d$ denote the momentum for the signal photon and driving laser, respectively. As a result, the directional driving field breaks the time-reversal symmetry and leads to non-reciprocal transmittance for the signal light: the co-propagating signal photons can be coherently coupled with phonons, while the coupling between counter-propagating photons and the phonon is negligible.

Next, we carry out a detailed theoretical derivation, considering that a spherical resonator in which one mechanical mode with a frequency $\omega_m$ is coupled to an optical mode with a frequency $\omega_c$. The resonator supports two degenerate counter-clockwise (CCW) and clockwise (CW) optical modes with the same frequency $\omega_c$. The full Hamiltonian, including external driving, is given by (($\hbar = 1$)):

$$H = \omega_c a_{\text{cw}}^\dagger a_{\text{cw}} + \omega_c a_{\text{ccw}}^\dagger a_{\text{ccw}} + \omega_m b^\dagger b + g_0 a_{\text{cw}}^\dagger a_{\text{cw}} \left(b + b^\dagger\right) + g_0 a_{\text{ccw}}^\dagger a_{\text{ccw}} \left(b + b^\dagger\right) + i\sqrt{\kappa_{in}}\epsilon_d \left(a_{\text{cw}}^\dagger e^{-i\omega_d t} - a_{\text{cw}} e^{i\omega_d t}\right). \tag{4.2}$$

where $a_{\text{cw}}$, $a_{\text{ccw}}$ and $b$ are annihilation operators of the two degenerate CW and CCW optical modes and the mechanical mode, respectively; $g_0$ is the single-photon coupling rate, which corresponds to the optical cavity frequency shift per phonon excitation; $\epsilon_d$ is the driving field of the mode $a_{\text{cw}}$ with a frequency $\omega_d$; and $\kappa_{in}$ is the microfiber to sphere coupling rate.

With a very weak coupling rate $g_0 \ll \omega_m, \kappa$, where $\kappa$ cavity energy decay rate in the experiments of this study, the interplay between single photons in the CCW and CW optical modes that are mediated by a phonon can be neglected. To increase optomechanical interactions, the cavity mode $a_{\text{cw}}$ is driven by the driven mode using an external laser. In the frame rotation of $H_0 = \omega_d a_{\text{cw}}^\dagger a_{\text{cw}}$, we can make a unitary transformation $a_{\text{cw}} \rightarrow a_{\text{cw}} + \alpha_{\text{cw}}$. In these experiments, the duration of the driving pulse $\tau_d \gg 1/\kappa$; thus, we can approximately treat the driving field as a constant field:

$$\alpha_{\text{cw}} = \frac{\sqrt{\kappa_{in}}\epsilon_d}{-i\Delta_d + \kappa/2}, \tag{4.3}$$

where the detuning $\Delta_d = \omega_d - \omega_c$.

By substituting a complex field amplitude into the Hamiltonian, the nonlinear optomechanical coupling is linearized in the resolved-sideband limit, which is given by:

$$H_{lin} = -\Delta_d a_{\text{cw}}^\dagger a_{\text{cw}} + \omega_c a_{\text{ccw}}^\dagger a_{\text{ccw}} + \omega_m b^\dagger b + \left(G_{\text{cw}} a_{\text{cw}}^\dagger + G_{\text{cw}}^* a_{\text{cw}}\right)\left(b + b^\dagger\right), \tag{4.4}$$

where $G_{\text{cw}} = g_0\alpha_{\text{cw}}$ is the enhanced optomechanical coupling strength. We have thus neglected the term $g_0 a_{\text{ccw}}^\dagger a_{\text{ccw}} \left(b + b^\dagger\right)$, which leads to an asymmetry between the CCW and CW modes, allowing optical non-reciprocal effects.

When signal light is input into the system, the Hamiltonian becomes:

$$
\begin{aligned}
H_{sys} =& -\Delta_d a_{\mathrm{cw}}^{\dagger} a_{\mathrm{cw}} + \omega_c a_{\mathrm{ccw}}^{\dagger} a_{\mathrm{ccw}} + \omega_m b^{\dagger} b + \left(G_{\mathrm{cw}} a_{\mathrm{cw}}^{\dagger} + G_{\mathrm{cw}}^{*} a_{\mathrm{cw}}\right)\left(b + b^{\dagger}\right) \\
&+ i\sqrt{\kappa_{in}}\epsilon_{s,\mathrm{cw}}\left(a_{\mathrm{cw}}^{\dagger} e^{-i(\omega_s-\omega_d)t} - a_{\mathrm{cw}} e^{i(\omega_s-\omega_d)t}\right) \\
&+ i\sqrt{\kappa_{in}}\epsilon_{s,\mathrm{ccw}}\left(a_{\mathrm{ccw}}^{\dagger} e^{-i\omega_s t} - a_{\mathrm{ccw}} e^{i\omega_s t}\right), \qquad (4.5)
\end{aligned}
$$

where $\epsilon_{s,\mathrm{cw}}$ and $\epsilon_{s,\mathrm{ccw}}$ represent CW and CCW optical modes with a frequency $\omega_s$. Only the CW optical mode excited by the signal fields is related to the mechanical mode, and the CCW optical mode evolves independently.

### 4.2.1 *Red-Detuned Driving*

When the mode $a_{\mathrm{cw}}$ is driven at the mechanical red sideband ($\Delta_d = -\omega_m$), in the rotating frame of Hamiltonian $H_0 = (\omega_s - \omega_d)\, a_{\mathrm{cw}}^{\dagger} a_{\mathrm{cw}} + \omega_s a_{\mathrm{ccw}}^{\dagger} a_{\mathrm{ccw}} + (\omega_s - \omega_d)\, b^{\dagger} b$, the effective Hamiltonian of the system can be obtained as:

$$
\begin{aligned}
H_{eff} =& -\delta a_{\mathrm{cw}}^{\dagger} a_{\mathrm{cw}} - \delta a_{\mathrm{ccw}}^{\dagger} a_{\mathrm{ccw}} - \Delta_m b^{\dagger} b + G_{\mathrm{cw}} a_{\mathrm{cw}}^{\dagger} b + G_{\mathrm{cw}}^{*} a_{\mathrm{cw}} b^{\dagger} \\
&+ i\sqrt{\kappa_{in}}\epsilon_{s,\mathrm{cw}}\left(a_{\mathrm{cw}}^{\dagger} - a_{\mathrm{cw}}\right) + i\sqrt{\kappa_{in}}\epsilon_{s,\mathrm{ccw}}\left(a_{\mathrm{ccw}}^{\dagger} - a_{\mathrm{ccw}}\right), \qquad (4.6)
\end{aligned}
$$

where the detuning are $\delta = \omega_s - \omega_c$, $\Delta_m = \omega_s - \omega_d - \omega_m = \delta$.

When the driving light of mode $a_{\mathrm{cw}}$ is red detuned, and the detuned frequency is mechanical vibration frequency, that is ($\Delta_d = -\omega_m$), in the rotating reference frame of Hamiltonian $H_0 = (\omega_s - \omega_d)\, a_{\mathrm{cw}}^{\dagger} a_{\mathrm{cw}} + \omega_s a_{\mathrm{ccw}}^{\dagger} a_{\mathrm{ccw}} + (\omega_s - \omega_d)\, b^{\dagger} b$, the effective Hamiltonian of the system can be obtained as:

$$
\begin{aligned}
H_{eff} =& -\delta a_{\mathrm{cw}}^{\dagger} a_{\mathrm{cw}} - \delta a_{\mathrm{ccw}}^{\dagger} a_{\mathrm{ccw}} - \Delta_m b^{\dagger} b + G_{\mathrm{cw}} a_{\mathrm{cw}}^{\dagger} b + G_{\mathrm{cw}}^{*} a_{\mathrm{cw}} b^{\dagger} \\
&+ i\sqrt{\kappa_{in}}\epsilon_{s,\mathrm{cw}}\left(a_{\mathrm{cw}}^{\dagger} - a_{\mathrm{cw}}\right) + i\sqrt{\kappa_{in}}\epsilon_{s,\mathrm{ccw}}\left(a_{\mathrm{ccw}}^{\dagger} - a_{\mathrm{ccw}}\right), \qquad (4.7)
\end{aligned}
$$

where the detuning are $\delta = \omega_s - \omega_c$, $\Delta_m = \omega_s - \omega_d - \omega_m = \delta$.

In the steady state, we have: $\frac{da_{\mathrm{cw}}}{dt} = \frac{da_{\mathrm{ccw}}}{dt} = \frac{db}{dt} = 0$,

$$
a_{\mathrm{cw}}(\delta) = \frac{-\sqrt{\kappa_{in}}\epsilon_{s,\mathrm{cw}}}{i\delta - \frac{\kappa}{2}\left(1 + \frac{C_{\mathrm{cw}}}{1 - i\delta/\frac{\gamma_m}{2}}\right)}, \qquad (4.8)
$$

$$
a_{\mathrm{ccw}}(\delta) = \frac{-\sqrt{\kappa_{in}}\epsilon_{s,\mathrm{ccw}}}{i\delta - \frac{\kappa}{2}}, \qquad (4.9)
$$

$$
b(\delta) = \frac{iG_{\mathrm{ccw}}^{*} a_{\mathrm{ccw}}}{i\delta - \frac{\gamma_m}{2}}. \qquad (4.10)
$$

Here $C_{\mathrm{cw}} = \frac{4|G_{\mathrm{cw}}|^2}{\kappa\gamma_m}$ is the cooperativity.

### 4.2.2 Blue-Detuned Driving

When we choose the driving laser detuning at the blue sideband ($\Delta_d = \omega_m$), in the rotating frame of $H_0 = (\omega_s - \omega_d)\, a_{\text{cw}}^\dagger a_{\text{cw}} + \omega_s a_{\text{ccw}}^\dagger a_{\text{ccw}} - (\omega_s - \omega_d)\, b^\dagger b$, the effective Hamiltonian of the system is:

$$H_{eff} = -\,\delta a_{\text{cw}}^\dagger a_{\text{cw}} - \delta a_{\text{ccw}}^\dagger a_{\text{ccw}} - \Delta_m b^\dagger b + G_{\text{cw}} a_{\text{cw}}^\dagger b^\dagger + G_{\text{cw}}^* a_{\text{cw}} b$$
$$+\, i\sqrt{\kappa_{in}}\epsilon_{s,\text{cw}} \left(a_{\text{cw}}^\dagger - a_{\text{cw}}\right) + i\sqrt{\kappa_{in}}\epsilon_{s,\text{ccw}} \left(a_{\text{ccw}}^\dagger - a_{\text{ccw}}\right), \tag{4.11}$$

Where the detuning are rewritten as $\delta = \omega_s - \omega_c$, $\Delta_m = \omega_d - \omega_s - \omega_m = -\delta$.

The dynamics of the optical modes and mechanical mode are:

$$\frac{da_{\text{cw}}}{dt} = \left(i\delta - \frac{\kappa}{2}\right) a_{\text{cw}} - iG_{\text{cw}} b^\dagger + \sqrt{\kappa_{in}}\epsilon_{s,\text{cw}}, \tag{4.12}$$

$$\frac{da_{\text{ccw}}}{dt} = \left(i\delta - \frac{\kappa}{2}\right) a_{\text{ccw}} + \sqrt{\kappa_{in}}\epsilon_{s,\text{ccw}}, \tag{4.13}$$

$$\frac{db^\dagger}{dt} = \left(i\delta - \frac{\gamma_m}{2}\right) b^\dagger + iG_{\text{cw}}^* a_{\text{cw}}. \tag{4.14}$$

Similar to the red-detuned case, we have the steady-state spectrum:

$$a_{\text{cw}}(\delta) = \frac{-\sqrt{\kappa_{in}}\epsilon_{s,\text{cw}}}{i\delta - \frac{\kappa}{2}\left(1 + \frac{C_{\text{cw}}}{1 - i\delta/\frac{\gamma_m}{2}}\right)}, \tag{4.15}$$

$$a_{\text{ccw}}(\delta) = \frac{-\sqrt{\kappa_{in}}\epsilon_{s,\text{ccw}}}{i\delta - \frac{\kappa}{2}}, \tag{4.16}$$

$$b^\dagger(\delta) = \frac{-iG_{\text{cw}}^* a_{\text{cw}}}{i\delta - \frac{\gamma_m}{2}}. \tag{4.17}$$

here, $C_{\text{cw}} = -\frac{4|G_{\text{cw}}|^2}{\kappa\gamma_m}$.

When $C_{\text{cw}} \to -1$, $|a_{\text{cw}}| \to \infty$ leads to instability of the system. Therefore, we solve the system evolution with finite interaction time $t$, then obtain

$$a_{\text{cw}}(\delta) = a_+ e^{P_+ t} + a_- e^{P_- t} + a_0, \tag{4.18}$$

where

$$P_\pm = \frac{\left(2i\delta - \frac{\kappa+\gamma_m}{2}\right) \pm \sqrt{\left(\frac{\kappa-\gamma_m}{2}\right)^2 + 4\left|G_{\text{cw}}\right|^2}}{2} \tag{4.19}$$

$$a_+ = \frac{\sqrt{\kappa_{in}}\epsilon_{s,\text{cw}}\left[P_- - \left(i\delta - \frac{\kappa}{2}\right)\right]}{P_+\left(P_- - P_+\right)} \tag{4.20}$$

$$a_{-} = \frac{\sqrt{\kappa_{in}}\epsilon_{s,\mathrm{cw}}\left[P_{+} - \left(i\delta - \frac{\kappa}{2}\right)\right]}{P_{-}\left(P_{+} - P_{-}\right)} \tag{4.21}$$

$$a_{0} = \frac{-\sqrt{\kappa_{in}}\epsilon_{s,\mathrm{cw}}\left[\left(P_{+} + P_{-}\right) - \left(i\delta - \frac{\kappa}{2}\right)\right]}{P_{+}P_{-}} \tag{4.22}$$

when $\kappa \gg \gamma_m, \ |G_{\mathrm{cw}}|$, we can simplify the above equation as:

$$a_{\mathrm{cw}}(\delta) \approx \frac{-\sqrt{\kappa_{in}}\epsilon_{s,\mathrm{cw}}}{i\delta - \frac{\kappa}{2}\left(1 + \frac{C_{\mathrm{cw}}}{1 - i\delta/\frac{\gamma_m}{2}}\right)} \times \left\{1 + \frac{2C_{\mathrm{cw}}e^{\left(i\delta - \frac{\kappa+\gamma_m}{4}\right)t}}{1 - i\delta/\frac{\gamma_m}{2}} \sinh\left[\frac{t}{2}\sqrt{\left(\frac{\kappa+\gamma_m}{2}\right)^2 - \kappa\gamma_m\left(1 + C_{\mathrm{cw}}\right)}\right]\right\}. \tag{4.23}$$

### 4.2.3 *Two Red-Detuned Driving*

When we have two driving fields on both CCW and CW optical modes, the linearized Hamiltonian is given by:

$$\begin{aligned} H_{lin} =& -\Delta_{d,\mathrm{ccw}}a_{\mathrm{ccw}}^{\dagger}a_{\mathrm{ccw}} - \Delta_{d,\mathrm{cw}}a_{\mathrm{cw}}^{\dagger}a_{\mathrm{cw}} + \omega_m b^{\dagger}b + \left(G_{\mathrm{ccw}}a_{\mathrm{ccw}}^{\dagger} + G_{\mathrm{ccw}}^{*}a_{\mathrm{ccw}}\right)\left(b + b^{\dagger}\right) \\ & + \left(G_{\mathrm{cw}}a_{\mathrm{cw}}^{\dagger} + G_{\mathrm{cw}}^{*}a_{\mathrm{cw}}\right)\left(b + b^{\dagger}\right), \end{aligned} \tag{4.24}$$

where $\Delta_{d,\mathrm{ccw}} = \omega_{d,\mathrm{ccw}} - \omega_c$, $\left(\Delta_{d,\mathrm{cw}} = \omega_{d,\mathrm{cw}} - \omega_c\right)$ with $\omega_{d,\mathrm{ccw}}$ is the frequency of the CCW driving field, $G_{\mathrm{ccw}} = g_0\alpha_{\mathrm{ccw}}$ and $G_{\mathrm{cw}} = g_0\alpha_{\mathrm{cw}}$ are the effective optomechanical coupling rates, in which the classical field amplitudes of the driven modes are:

$$\alpha_{\mathrm{cw}} = \frac{\sqrt{\kappa_{in}}\epsilon_{d,\mathrm{cw}}}{-i\Delta_{d,\mathrm{cw}} + \kappa/2}, \tag{4.25}$$

$$\alpha_{\mathrm{ccw}} = \frac{\sqrt{\kappa_{in}}\epsilon_{d,\mathrm{ccw}}}{-i\Delta_{d,\mathrm{ccw}} + \kappa/2}. \tag{4.26}$$

Here, $\epsilon_{d,\mathrm{cw}}$ and $\epsilon_{d,\mathrm{ccw}}$ denote the driving field of the driven CCW and CW optical modes, respectively.

For studying the optical mode conversion of the optomechanical system, we just probe the system by CW signal field. The Hamiltonian of the system can be described by

$$
\begin{aligned}
H_{sys} = & -\Delta_{d,\text{ccw}} a_{\text{ccw}}^{\dagger} a_{\text{ccw}} - \Delta_{d,\text{cw}} a_{\text{cw}}^{\dagger} a_{\text{cw}} + \omega_m b^{\dagger} b + \left( G_{\text{ccw}} a_{\text{ccw}}^{\dagger} + G_{\text{ccw}}^{*} a_{\text{ccw}} \right) \left( b + b^{\dagger} \right) \\
& + \left( G_{\text{cw}} a_{\text{cw}}^{\dagger} + G_{\text{cw}}^{*} a_{\text{cw}} \right) \left( b + b^{\dagger} \right) \\
& + i \sqrt{\kappa_{in}} \epsilon_{s,\text{cw}} \left( a_{\text{cw}}^{\dagger} e^{-i(\omega_s - \omega_{d,\text{cw}})t} - a_r e^{i(\omega_s - \omega_{d,\text{cw}})t} \right).
\end{aligned} \tag{4.27}
$$

For the red detuning $\Delta_{d,\text{ccw}} = \Delta_{d,\text{cw}} = -\omega_m$, we can obtain the effective Hamiltonian as

$$
\begin{aligned}
H_{eff} = & -\delta_{\text{ccw}} a_{\text{ccw}}^{\dagger} a_{\text{ccw}} - \delta_{\text{cw}} a_{\text{cw}}^{\dagger} a_{\text{cw}} - \Delta_m b^{\dagger} b + G_{\text{ccw}} a_{\text{ccw}}^{\dagger} b + G_{\text{ccw}}^{*} a_{\text{ccw}} b^{\dagger} \\
& + G_{\text{cw}} a_{\text{cw}}^{\dagger} b + G_{\text{cw}}^{*} a_{\text{cw}} b^{\dagger} + i \sqrt{\kappa_{in}} \epsilon_{s,\text{cw}} \left( a_{\text{cw}}^{\dagger} - a_{\text{cw}} \right),
\end{aligned} \tag{4.28}
$$

where $\delta_{\text{cw}} = \omega_s - \omega_c, \delta_{\text{ccw}} = \omega_s - \omega_c + \omega_{d,\text{ccw}} - \omega_{d,\text{cw}} = \delta_{\text{cw}}, \Delta_m = \omega_s - \omega_{d,\text{cw}} - \omega_m = \delta_{\text{cw}}$.

The dynamics of the optical modes and mechanical mode are

$$
\frac{da_{\text{cw}}}{dt} = (i\delta_{\text{cw}} - \kappa/2)\, a_{\text{cw}} - iG_{\text{cw}} b + \sqrt{\kappa_{in}} \epsilon_{s,\text{cw}}, \tag{4.29}
$$

$$
\frac{da_{\text{ccw}}}{dt} = (i\delta_{\text{cw}} - \kappa/2)\, a_{\text{ccw}} - iG_{\text{ccw}} b, \tag{4.30}
$$

$$
\frac{db}{dt} = (i\delta_{\text{cw}} - \gamma_m/2)\, b - iG_{\text{ccw}}^{*} a_{\text{ccw}} - iG_{\text{cw}}^{*} a_{\text{cw}}. \tag{4.31}
$$

We have the steady state solution of optical cavity field as

$$
a_{\text{cw}} = \frac{-\sqrt{\kappa_{in}} \epsilon_{s,\text{cw}}}{i\delta_{\text{cw}} - \frac{\kappa}{2} \left( 1 + \frac{C_{\text{cw}}}{1 - 2i\delta_{\text{cw}}/\gamma_m + C_{\text{ccw}}/(1 - 2i\delta_{\text{cw}}/\kappa)} \right)}, \tag{4.32}
$$

$$
a_{\text{ccw}} = \frac{4G_{\text{ccw}} G_{\text{cw}}^{*} \sqrt{\kappa_{in}} \epsilon_{s,\text{cw}} / (\kappa \gamma_m)}{(i\delta_{\text{cw}} - \kappa/2) \left[ (1 - 2i\delta_{\text{cw}}/\gamma_m)(1 - 2i\delta_{\text{cw}}/\kappa) + C_{\text{ccw}} \right] + (i\delta_{\text{cw}} - \kappa/2)\, C_{\text{cw}}}. \tag{4.33}
$$

Specially, the photon mode-conversion efficiency at $\delta_{\text{cw}} = 0$ is given by:

$$
\eta = \frac{\kappa_{in} \left| a_{\text{ccw}} \right|^2}{\epsilon_{s,\text{cw}}^2} = \left( \frac{\kappa_{in}}{\kappa} \right)^2 \frac{4C_{\text{ccw}} C_{\text{cw}}}{(1 + C_{\text{ccw}} + C_{\text{cw}})^2}. \tag{4.34}
$$

Therefore, high efficiency optical conversion from the CW optical mode to CCW optical mode mediated by phonon is possible.

The experimental demonstration of the optomechanically induced non-reciprocity is performed using the experimental setup shown in Fig. 4.3a. The driving laser excites the CW optical mode, while either CW or CCW signal photons are sent to the cavity. In a sphere with a diameter of approximately 36 μm, we choose a high quality factor WGM near 780 nm with linewidth of $\kappa/2\pi = 15$ MHz. The radial

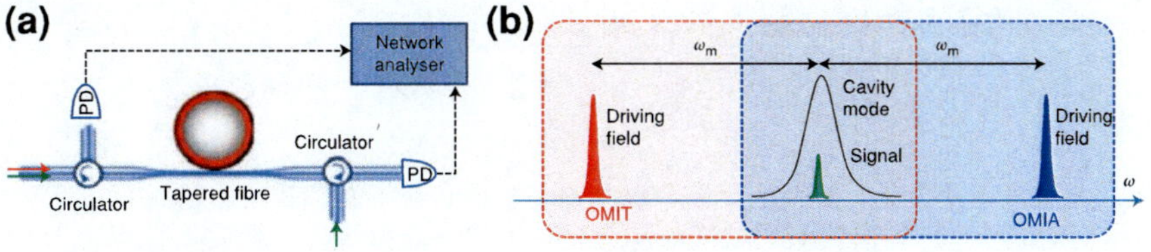


**Fig. 4.3** **a** A simplified schematic of the experimental setup with a co- or counter-propagating signal in a silica microsphere. PD, photodetector. **b** Spectral position of the driving and signal field in OMIT and OMIA processes

**Fig. 4.4** **a–b** The emission power spectra in the OMIT response are obtained using a co- and counter-propagating signal pulse. The incident driving power is 10 mW. The solid red lines are the results of calculations using the parameters $\omega_m/2\pi = 88.54$ MHz, $\kappa/2\pi = 15$ MHz, $\gamma_m/2\pi = 22$ kHz, $C_{\rm cw} = 1.5$

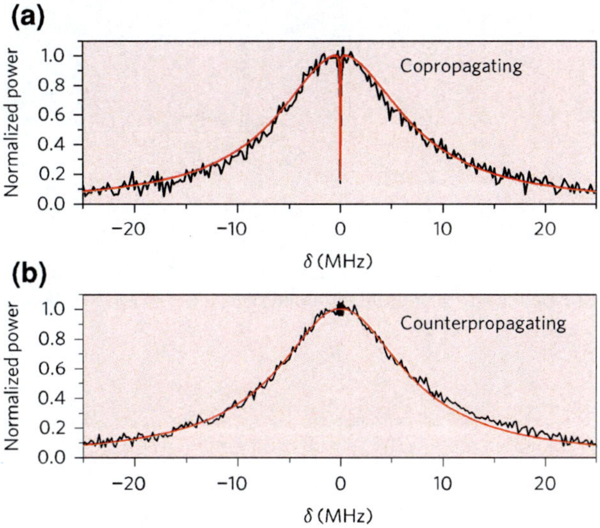


breathing mechanical mode in this sphere has a frequency of $\omega_m/2\pi = 88.54$ MHz and a linewidth of $\gamma_m/2\pi = 22$ kHz.

Figure 4.3b shows the two configurations of the driving field. For the driving field that is red-detuned from the cavity mode by $\omega_m$, an effective photon-phonon beamsplitter-like interaction $(a_{\rm cw}^{\dagger}b + a_{\rm cw}b^{\dagger})$ would lead to coherent conversion and induce a transparent window in the cavity field spectrum [19, 20]. In this experiment, we send a short drive laser and signal pulse (pulse duration $\tau_p = 18\,\mu$s) to measure the transient optomechanical coupling to avoid thermal instability in the microsphere [21]. The experimental result in Fig. 4.4a shows a sharp transparency window for co-propagating signal photons, demonstrating the destructive interference between the signal field and the optical field generated from the anti-Stokes scattering process. Conversely, the spectrum of the counterpropagating signal in Fig. 4.4b does not show such an effect, which is indicative of optomechanically induced non-reciprocity. The spectra still agree with the theoretical predictions of the steady-state intracavity field:

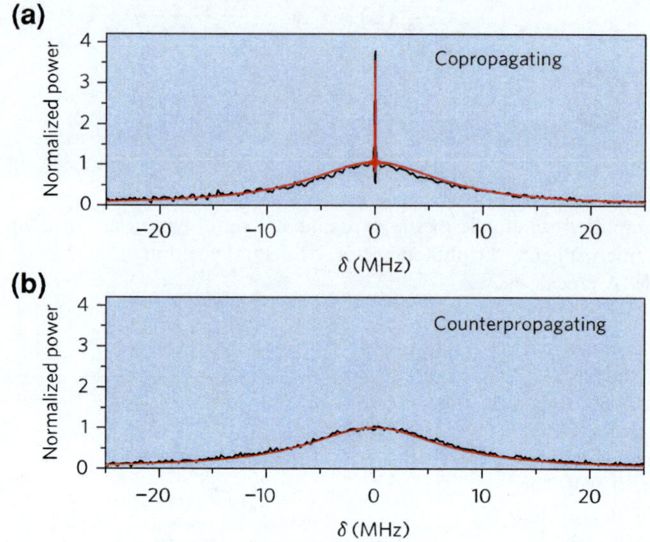


**Fig. 4.5 a–b** In the OMIA process, the emission power spectra are obtained using a co- and counter-propagating signal pulse. The incident driving power is 5.5 mW. The solid red lines are the results of calculations using the parameters $\omega_m/2\pi = 88.54$ MHz, $\kappa/2\pi = 15$ MHz, $\gamma_m/2\pi = 22$ kHz, and $C_{\rm cw} = -0.8$

$$a_{\rm cw}(\delta) = \frac{-\sqrt{\kappa_{in}}\epsilon_{s,\rm cw}}{i\delta - \frac{\kappa}{2}\left(1 + \frac{C_{\rm cw}}{1-i2\delta/\gamma_m}\right)}, \tag{4.35}$$

$$a_{\rm ccw}(\delta) = \frac{-\sqrt{\kappa_{in}}\epsilon_{s,\rm ccw}}{i\delta - \frac{\kappa}{2}}. \tag{4.36}$$

where $C_{\rm cw} = \frac{4|G_{\rm cw}|^2}{\kappa\gamma_m}$ is the cooperativity; $G_{\rm cw} = \sqrt{N_d}g_0$ is the effective optomechanical coupling rate; $N_d$ is the CW driving intracavity photon number; $\epsilon_{s,\rm cw}$ and $\epsilon_{s,\rm ccw}$ are the weak signal amplitudes of the CW- and CCW-circulating optical modes, respectively; $\delta$ is the detuning between the signal and cavity field; and $\kappa_{in}$ is the coupling rate between the microfiber and microsphere. For a critically coupled optical mode, the observed intracavity fields for the CW and CCW signal corresponding to non-reciprocal transmissions of the CW and CCW signal are 0 and 1 around $\delta \sim 0$, respectively.

For the driving field that is blue-detuned from the cavity mode by $\omega_m$, an effective photon-phonon pair generation process $(a^\dagger_{\rm cw}b^\dagger + a_{\rm cw}b)$ would lead to optomechanically induced amplification. Figure 4.5a, b shows the corresponding experimental results, which show non-reciprocal amplification of the signal. The sequences of driving laser and signal pulses are the same as before. Here, the experimental results are fitted by the modified transient intracavity field spectra because of the unsteady state optomechanical coupling even with 18 μs driving pulse at blue sideband:

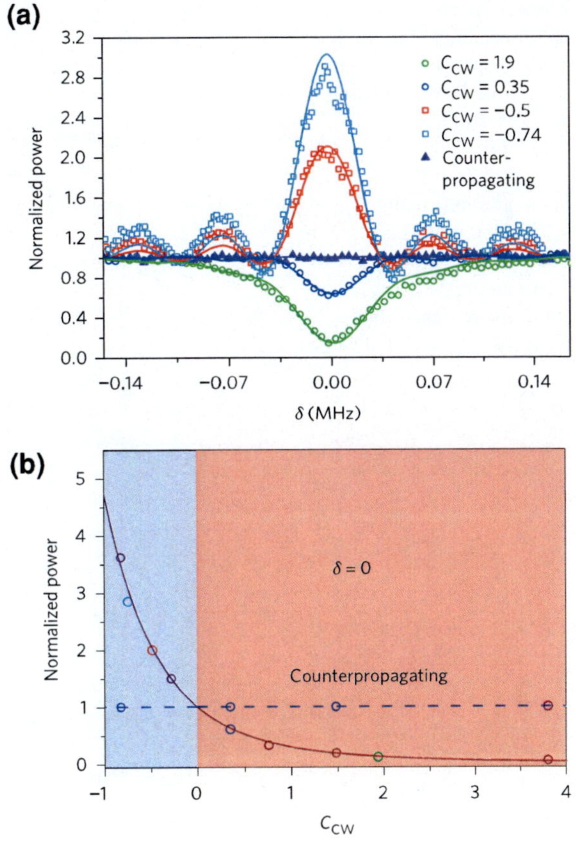


**Fig. 4.6** **a** The typical emission power from the microcavity with the co-propagating (circle for OMIT and square for OMIA) and counter-propagating signal (triangle). The solid lines are the results of calculations using $C_{\text{cw}} = -0.74, -0.5, 0, 0.35, 1.9$. **b** The spectral depth of the transparency window or spectral peak of the opacity peak as a function of the cooperativity $C_{\text{cw}}$

$$a_{\text{cw}}(\delta) = \frac{-\sqrt{\kappa_{in}}\epsilon_{s,\text{cw}}}{i\delta - \frac{\kappa}{2}\left(1 + \frac{C_{\text{cw}}}{1-i2\delta/\gamma_m}\right)}[1 + F(\tau_p)], \tag{4.37}$$

with the transient modification $F(t) = \frac{2C_{\text{cw}}Exp(i\delta t-(\kappa+\gamma_m)t/4)}{1-i2\delta/\gamma_m}\sinh\left[\frac{t\sqrt{(\kappa+\gamma_m)^2/4-\kappa\gamma_m(1+C_{\text{cw}})}}{2}\right]$. Here, $C_{\text{cw}} = -\frac{4|G_{\text{cw}}|^2}{\kappa\gamma_m}$ has a negative sign, which would effectively narrow the linewidth of the cavity and amplify the intracavity signal.

To obtain a more quantitative analysis of the optomechanically induced non-reciprocity, the detailed spectra for different experimental conditions are shown in Fig. 4.6a. For blue detuning, the peak height increases and the linewidth decreases with the driving power. For red detuning, the transparent depth and linewidth both increase with the driving power. The intensity at $\delta = 0$ and the corresponding cooperativity $C_{\text{cw}}$ are summarized and plotted in Fig. 4.6b. The intensity of the co-propagating signal agrees with the prediction, while the counter-propagating signal is found to be independent of $C_{\text{cw}}$.

The non-reciprocal transmittance of the signal can also be inferred from the phase because $\arg(a_{\text{cw}}(\delta))$ in Eq. (4.35) varies with the detuning and $C_{\text{cw}}$. The measured

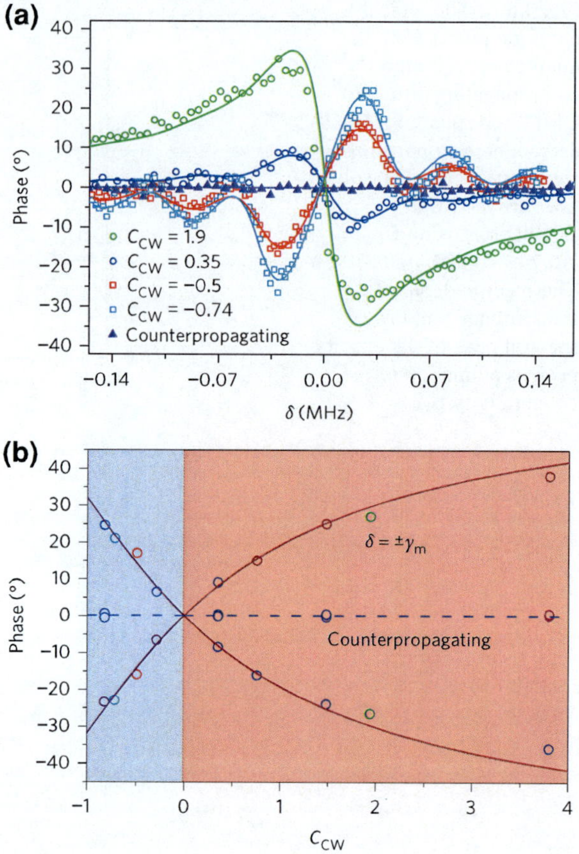


**Fig. 4.7** **a** The optomechanical interaction phase response (circle and square for co-propagating signal, triangle for counter-propagating signal). The solid lines are the results of calculations using $C_{\mathrm{cw}} = -0.74, -0.5, 0, 0.35, 1.9$. **b** The shift phase obtained at the detuning $\delta = \pm\gamma_m$. The solid line represent the theoretically expected values

phase is plotted in Fig. 4.7a. The difference between the co-propagating and counter-propagating signal unambiguously shows the non-reciprocal phase shifter due to the optomechanical interaction. The extracted phases at $\delta = \pm\gamma_m$ that are plotted in Fig. 4.7b agree with the theory. The maximum phase shift achieved in the proposed experiment is approximately 40°. The oscillations around the opacity peak in Fig. 4.6a and the corresponding phase response in Fig. 4.7a are observed due to the transient response of the optomechanical system, as described by $F(t)$ [22].

Optomechanically induced non-reciprocity is actually controllable using two oppositely propagating driving fields that excite the CW and CCW modes simultaneously. As a result of the interesting interplay between the three coupled modes (Fig. 4.8), there is an optomechanically dark mode [23] (a superposition of the CW and CCW modes), which enables the conversion of optical fields from the CW mode to the CCW mode. To test the optical mode conversion, we fixed the CW driving power at 10 mW and varied the CCW driving power from 0 to 18 mW, while the input signal coupled only to the CW mode. As shown in Fig. 4.9, the excitation of the CW mode increases with increasing $C_{\mathrm{ccw}}$, and the OMIT dip for this mode vanishes and is accompanied by a spectral broadening of the dip. There is light

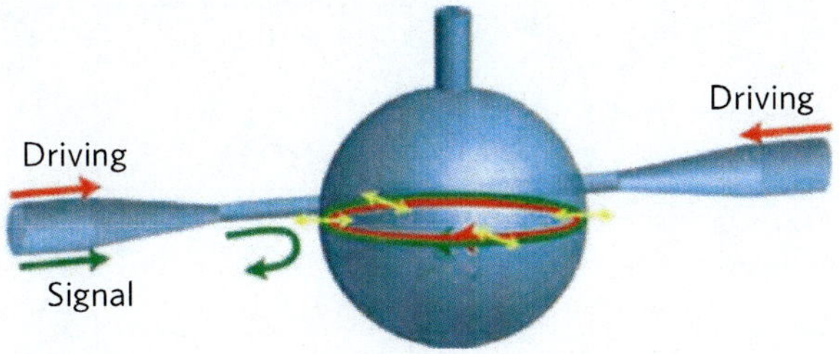


**Fig. 4.8** Two driving fields are applied simultaneously to induce the reflection of the signal

reflected from the system with a bandwidth smaller than 100 kHz; its peak power at $\delta \sim 0$ also increases with $C_{\mathrm{ccw}}$. The theoretical model predicts that $a_{\mathrm{cw}} \propto \frac{1+C_{\mathrm{ccw}}}{1+C_{\mathrm{cw}}+C_{\mathrm{ccw}}}$ and $a_{\mathrm{ccw}} \propto \frac{\sqrt{C_{\mathrm{cw}}C_{\mathrm{ccw}}}}{1+C_{\mathrm{cw}}+C_{\mathrm{ccw}}}$ for the CW signal, which agrees with the experiment results shown in Fig. 4.10. Compared to the CCW signal, for which $a_{\mathrm{ccw}} \propto \frac{1+C_{\mathrm{cw}}}{1+C_{\mathrm{cw}}+C_{\mathrm{ccw}}}$ and $a_{\mathrm{cw}} \propto \frac{\sqrt{C_{\mathrm{cw}}C_{\mathrm{ccw}}}}{1+C_{\mathrm{cw}}+C_{\mathrm{ccw}}}$, the co-propagating intracavity excitation and transmission are non-reciprocal if $C_{\mathrm{CW}} \neq C_{\mathrm{CCW}}$, while the signal reflection is always reciprocal. Therefore, the system behaves as a controllable narrowband reflector with nonreciprocal transmittance, which might be interesting for future studies.

## 4.3 Experimental Setup

To measure the non-reciprocal responses of the optomechanical system, the transmission of the signal must be measured in different propagation directions. Therefore, we should be able to control the direction of the signal to excite either the CW or CCW traveling-wave optical mode. To accomplish that, we generate the signal light by modulating a weak probe beam of around 50 μW, which is markedly detuned from the optical mode. The blue sideband, which is nearly resonant with the optical mode, serves as the signal light. And the propagation direction of the signal can be switched. In Fig. 4.11a, b, the spectral positions of the driving, weak probe beam and signal lights used in the experiment are shown. To avoid effects from the probe beam and the red-sideband of the probe beam, the frequency of the probe beam is 200 MHz away from the cavity resonance.

Figure 4.11c shows the schematic of the pulse sequence and the detection gate used for these experiments. The driving and signal light rectangle pulses are synchronized. In each repeated experiment, detection is followed by a procedure of laser locking to the cavity mode.

In these experiments, a silica microsphere with a diameter of approximately 36 μm is used as the optomechanical system. Figure 4.12a shows the transmission spectrum of the selected high-quality-factor whispering gallery mode (WGM) with a wavelength of approximately 780 nm, which indicates $\kappa_{in} = \kappa/2 = 7.5 \times 2\pi$ MHz in the critical coupling regime. From the fitting of the spectrum by optimizing the squared

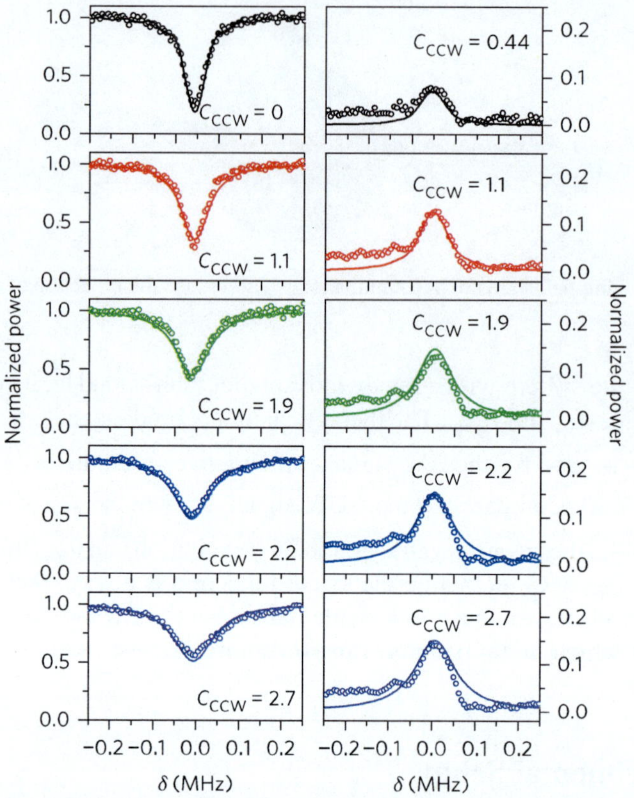


**Fig. 4.9** The cavity emission power spectra in the CW direction and the reflected power spectra in the CCW direction vary with the cooperativity $C_{ccw}$. The peak power of the CW driving pulse is 10 mW, corresponding to $C_{cw} = 1.5$. The solid lines are the theoretical calculation results

residuals between experiment results and fitting, we found that the backscattering induced coupling between left- and right-hand circulating mode $2\beta/\kappa \leq 0.1$. The effects of such small backscattering on the observed optomechanically induced transparency can be neglected. Figure 4.12b shows the displacement power spectrum of the radial breathing mode of the microsphere.

Figure 4.13 shows a detailed schematic of the experimental setup for the demonstration of the optomechanically induced non-reciprocity. As shown in Fig. 4.13, the strong driving pulse and the weak locking pulse were derived from the first-order diffraction of the laser, which passed through an acoustic-optic modulator ($AOM_1$). To avoid saturating the detector, the strong driving pulse was attenuated before being detected by the photo detector ($D_1$). A weak locking pulse should be maintained to ensure a good error signal ($D_2$); therefore, $AOM_3$ was introduced to separate these two pulses [21]. The error signal could be obtained for the proportional-integral-derivative (PID) controller to lock the laser at the relevant frequency using the Pound–Drever–Hall technique when the weak locking pulse propagated through an electro-optic modulator ($EOM_1$) with a modulation frequency $f = \omega_m$. The weak probe beam

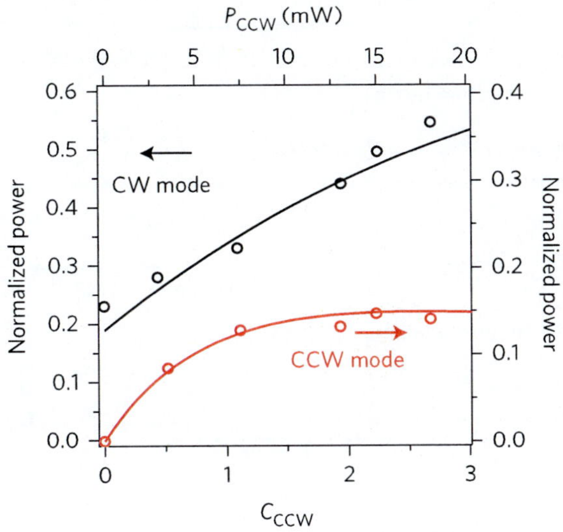


**Fig. 4.10** Emission powers from the CW mode (black circles) and CCW mode (red circles) at $\delta = 0$ as a function of $C_{ccw}$. The solid lines are results of the theoretical calculations

was generated by a separate laser beam from the same laser and was derived from $AOM_2$, the frequency of which was chosen to match the spectral position shown in Fig. 4.11 for a different experiment. When the probe beam passed through $EOM_2$, one sideband generated from $EOM_2$ served as the signal pulse and was nearly resonant with the optical modes. Thus, it was possible to launch the signal in either the left- or right-propagating directions using double circulators and a flip mirror.

For the optomechanically induced transparency (OMIT) process, the driving laser was locked at one frequency, $\omega_m$, below the given WGM resonance. For the optomechanically induced amplification (OMIA) process, the driving laser was locked to be blue-detuned from the optical mode by $\omega_m$.

In the experiment, both the driving and signal pulses were square-shaped. The pulses were synchronized and had the same duration of 18 $\mu$s. Heterodyne detection was used to measure the light emission from the microsphere with the weak probe pulse serving as the local oscillator. Because the incident signal pulse was generated directly from the local oscillator with a pure phase modulation, heterodyne detection was not sufficiently sensitive to the transmitted signal pulses that were not emitted from the optical resonator. Thus, the intracavity optical field was directly measured via heterodyne detection. The transient behavior of the optical emission was probed using a network analyzer that was operating in a time-gated detection mode with a resolution bandwidth of 1 MHz. The measured signal emission power spectra should be approximately 200 MHz. To facilitate understanding, the measured power spectra are shown for $\delta \sim 0$.

In optical mode conversion measurements, the signal that is created due to the interference of the reflected signal with the CCW driving pulse is described by using a network analyzer. Considering that a voltage signal converted through an

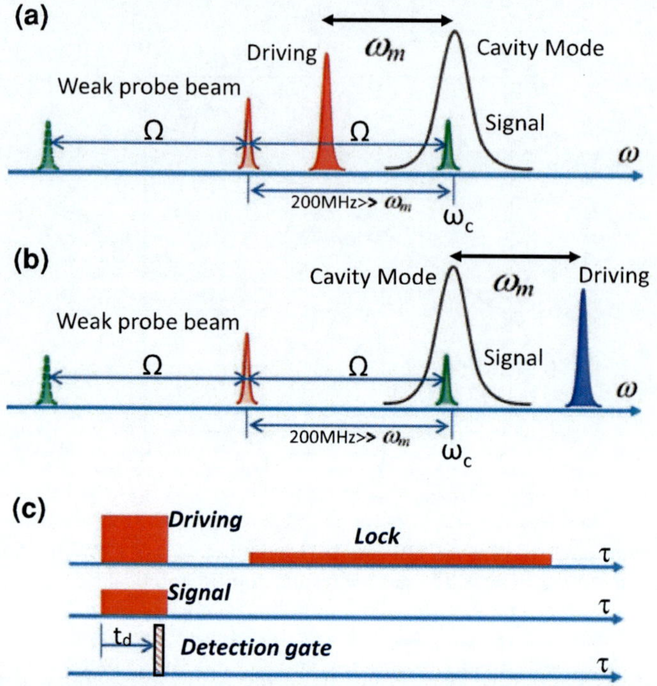


**Fig. 4.11** **a** In the OMIT experiment, the driving field is locked at the frequency at which red detuning from the optical mode by one mechanical frequency ($\omega_m$) occurs. The weak probe beam's frequency is detuned from the optical mode by 200 MHz. The signal is generated by sending the weak probe beam through an EOM with modulation frequency Ω. **b** In the OMIA experiment, the driving field is locked at the frequency at which blue detuning from the optical mode by one mechanical frequency ($\omega_m$) occurs. **c** Schematic of the pulse sequences and the detection gate used for the transient experiments

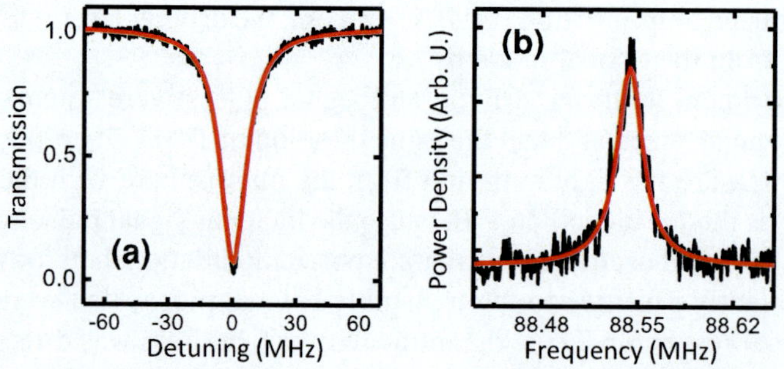


**Fig. 4.12** **a** Optical transmission spectrum for the optical mode in the microsphere. **b** Displacement power spectrum of the radial breathing mode obtained from the spectrum analyzer

AC-coupled photo-detector is proportional to $\sqrt{P_{reflect} P_{\text{ccw}}}$, where $P_{reflect}$ is the reflected signal power, and $P_{\text{ccw}}$ is the CCW pulse power, the network analyzer measures an RF power that is proportional to the square of this voltage, which is

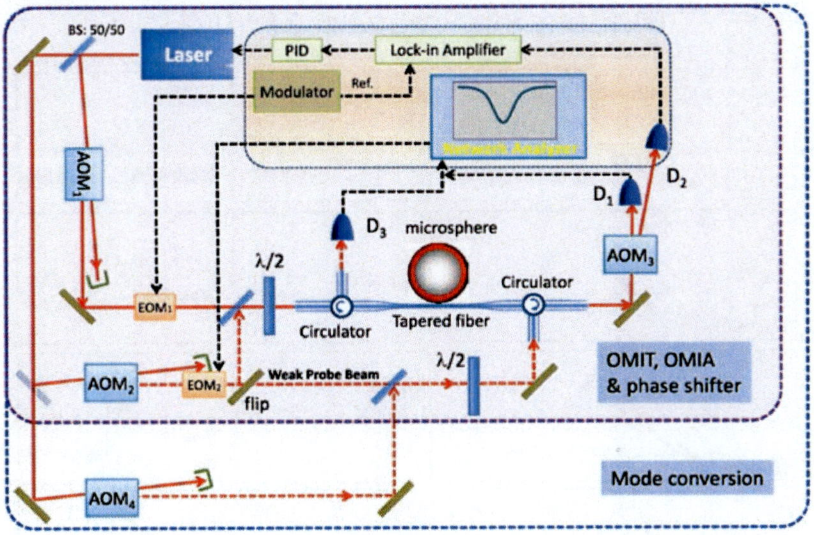


**Fig. 4.13** Detailed experimental setup. The red lines represent optical paths, and black lines represent electrical connections. AOM, EOM, PID, D, and $\lambda/2$ represent the acoustic-optic modulator, the electro-optic modulator, the proportional-integral-derivative controller, the detector, and the half-wave plate, respectively

also proportional to $P_{reflect} P_{\text{ccw}}$ [24]. The reflected photon flux can be obtained using the well-known input-output relation, which relates the intra-cavity field $\alpha_{\text{ccw}}$. Therefore, the RF power at $\omega_m$ in the counter-propagating direction is measured as $S_{\text{ccw}} = \xi_{\text{ccw}} P_{reflect} P_{\text{ccw}} = \xi_{\text{ccw}} k_{in} \mid \alpha_{\text{ccw}} \mid^2 P_{\text{ccw}}$, where $\xi_{\text{ccw}}$ is a constant that depends on the photo-detector gain and insertion or absorption losses in the optical path. The RF power at $\omega_m$ in the other direction is measured as $S_{\text{cw}} = \xi_{\text{cw}} \kappa_{in} \mid \alpha_{\text{cw}} \mid^2 P_{\text{cw}}$, where $P_{\text{cw}}$ is the CW pulse power. The input signal $P_{signal}$ can be estimated from the intracavity field $\alpha_{\text{cw}}$ with $P_{\text{ccw}} = 0$. Therefore, the conversion efficiency can easily be written as $\eta = P_{reflect}/P_{signal} = \frac{4k_{in}^2}{k^2(1+C_{\text{cw}})^2} |\alpha_{\text{ccw}}|^2 / |\alpha_{\text{cw}}|^2 = \frac{4k_{in}^2}{k^2(1+C_{\text{cw}})^2} S_{\text{ccw}} \xi_{\text{cw}} P_{\text{cw}} / (S_{\text{cw}} \xi_{\text{ccw}} P_{\text{ccw}})$. The efficiency can thus be determined from the RF spectra and the power of right- and left-propagating pulses.

As shown in Fig. 4.14, we also investigated the dependence of the OMIT and corresponding phase shift on the detuning $\Delta_d$ between the cavity resonance and the driving field with a fixed power of 10 mW. Although the peak of the cavity resonance shifts with $\Delta_d$, the transparency window and the phase shift were observed only when the two-photon detuning was near the mechanical resonance frequency ($\omega_s - \omega_d \approx \omega_m$), independently of the detuning $\Delta_d$ of the driving field.

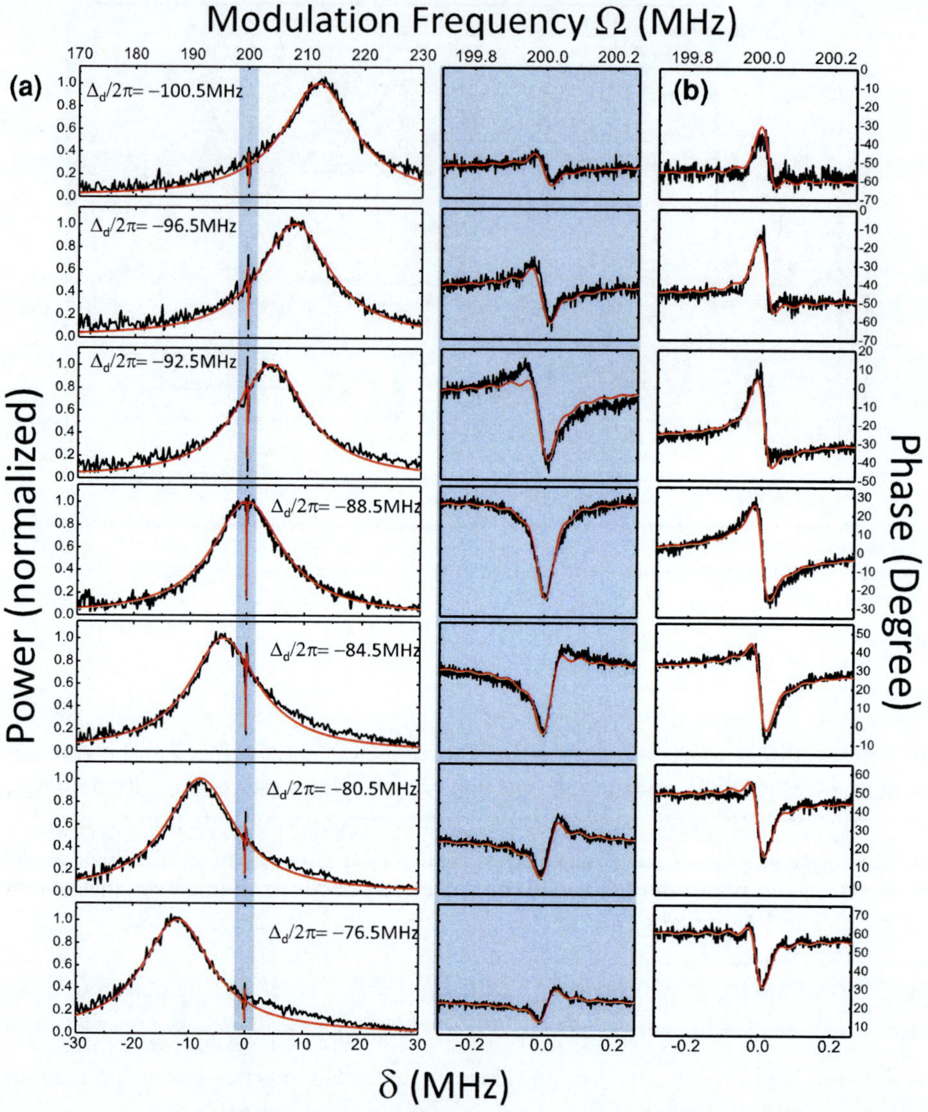


**Fig. 4.14** **a** Experimentally observed emission power when the signal frequency was scanned by sweeping the phase modulator frequency $\Omega$ for different values of driving beam detuning, $\Delta_d/2\pi = -100.5, -96.5, -92.5, -88.5, -84.5, -80.5, -76.5$ MHz. The incident driving power was 10 mW. The region around the central dip is shown in more detail with an expanded frequency scale. **b** The corresponding phase shift is also shown using the same frequency scale. While the center of the response of the bare optical cavity shifted with the driving field, the sharp dip of the OMIT and phase shift always occurred around the frequency $\Omega \sim 200$ MHz, corresponding to $\delta \sim 0$. Red lines in **a** and **b** are the theoretical calculations discussed

## 4.4 Summary

To summarize, optomechanically induced nonreciprocity is experimentally demonstrated for the first time in this study. The underlying mechanism of the non-reciprocity demonstrated in this study is actually universal and can be generalized to any traveling wave resonators via dispersive coupling with a mechanical resonator, such as the integrated ring-type microresonator coupled with a nanobeam [25]. Considering that higher cooperativity and cascading of the non-reciprocal devices are possible in a photonic integrated chip, optomechanically induced non-reciprocity has applications in integrated photonic isolators and circulators [16, 17], which will play important roles in a hybrid quantum Internet [26].

The main differences and advantages of optomechanical induced non-reciprocity with comparing the demonstrated Brillouin-scattering induced non-reciprocity are:

The Mode Matching Conditions

In general, the mode matching conditions of the optomechanically induced non-reciprocity can be more easily satisfied.

For Brillouin-scattering induced non-reciprocity, there are two optical modes and one mechanical mode involved. All the involved modes must satisfy the energy and momentum conservations. The driving laser is on-resonance with one of the optical modes, thus reduces the requirement of driving laser power. However, even in large-size spheres (diameter greater than 100 $\mu$m), only a handful of optical modes meet the mode matching conditions. Specially, it is hard to find two optical modes for the given Brillouin-scattering mode. For optomechanically induced non-reciprocity, there is only a single optical mode and a mechanical mode involved. In principle, the non-reciprocity can be achieved for optical modes in a broaden wavelength range, since the breath mechanical mode can change the radius of the sphere and thus couples to all optical modes. In our experiment, we just choose one typical optical mode, and the similar results can be obtained with the other optical modes with similar Q factor.

The Feasibility for Integration

The optomechanical systems based on traveling wave resonators have been widely used in integrated photonic chip, such as the GaAs disk resonator with waveguide [16], the microwheel resonator with waveguide [17], the racetrack microresonator with under-etching a part of waveguide [27], the integrated microresonator coupled with a nanomechanical oscillators [25].

Due to the mode matching conditions, it is almost impossible to stimulate Brillouin-scattering in these types of microresonators with a diameter of a few microns to tens of microns. The coherent Brillouin scattering has only been observed in the microsphere (diameter $\geq$ 100 microns) or microdisk (diameter $\approx$ 4 mm). For the first one, it is hard to be integrated in the photonic chips. For the microdisk case, the cascaded Brillouin scattering have been realized by matching the FSR of optical modes and the acoustic mode. However, a lot of optical modes be involved in this cascaded Brillouin scattering process, cascaded laser is demonstrated by the strong pump instead of the coherent coupling between photon and phonon.

In contrast, the underlying mechanism of optomechanically induced non-reciprocity demonstrated in our work can be easily generalized to these integrated traveling wave resonators. In addition, the mechanical mode not limited to the breath mode, it can be a uniform thickness mode, vibration of part of the resonator or a hybridized mechanical oscillator couple with the optical resonator through evanescent field. Therefore, the demonstrated optomechanical induced non-reciprocity is readily to be achieved in current integrated photonic chips using the microring or microdisk optical microresonators.

More Potential Applications

Because in traveling wave resonators, the clock-wise (CW) and counter-clockwise (CCW) optical modes can couple with the same mechanical mode, we can realize the controllable mode coupling between the CW and CCW traveling optical field. The demonstrated optomechanically induced non-reciprocity can also combine with the optomechanical frequency conversion, thus realize the devices that converted the photons propagating in certain direction to other wavelengths, while the photons propagating in the opposite direction remain the same. However, Brillouin-scattering induced non-reciprocity can't realize these applications, since the CW and CCW travelling mechanical modes in Brillouin-scattering are independently couple to CW and CCW optical fields. As a result, the Brillouin scattering can't induce the coupling between CW and CCW signals.

## References


1. Jalas D, Petrov A, Eich M, Freude W, Fan S, Yu Z, Baets R, Popovic M, Melloni A, Joannopoulos JD et al (2013) Nat Photonics 7(8):579
2. Bi L, Hu J, Jiang P, Kim DH, Dionne GF, Kimerling LC, Ross C (2011) Nat Photonics 5(12):758
3. Yu Z, Fan S (2009) Nat Photonics 3(2):91
4. Shi Y, Yu Z, Fan S (2015) Nat Photonics 9(6):388
5. Tzuang LD, Fang K, Nussenzveig P, Fan S, Lipson M (2014) Nat Photonics 8(9):701
6. Kang M, Butsch A, Russell PSJ (2011) Nat Photonics 5(9):549
7. Dong CH, Shen Z, Zou CL, Zhang YL, Fu W, Guo GC (2015) Nat Commun 6:6193
8. Kim J, Kuzyk MC, Han K, Wang H, Bahl G (2015) Nat Phys 11:275
9. Guo X, Zou CL, Jung H, Tang HX (2016) Phys Rev Lett 117(12):123902
10. Zheng Y, Yang J, Shen Z, Cao J, Chen X, Wan W (2016) Light: Sci Appl 5:e16072
11. Hua S, Wen J, Jiang X, Hua Q, Jiang L, Xiao M (2016) Nat Commun 7:13657
12. Hafezi M, Rabl P (2012) Opt Express 20(7):7672
13. Shen Z, Zhang YL, Chen Y, Zou CL, Xiao YF, Zou XB, Sun FW, Guo GC, Dong CH (2016) Nat Photonics 10:657
14. Ruesink F, Miri MA, Alu A, Verhagen E (2016) Nat Commun 7
15. Fang K, Luo J, Metelmann A, Matheny MH, Marquardt F, Clerk AA, Painter O (2017) Nat Phys 13:465
16. Gil-Santos E, Baker C, Nguyen D, Hease W, Gomez C, Lemaitre A, Ducci S, Leo G, Favero I (2015) Nat Nanotechnol 10(9):810

17. Fong KY, Poot M, Tang HX (2015) Nano Lett 15(9):6116
18. Park YS, Wang H (2009) Nat Phys 5:489
19. Weis S, Riviere R, Deleglise S, Gavartin E, Arcizet O, Schliesser A, Kippenberg TJ (2010) Science 330(6010):1520
20. Safavi-Naeini AH, Alegre TPM, Chan J, Eichenfield M, Winger M, Lin Q, Hill JT, Chang DE, Painter O (2011) Nature 472:69
21. Shen Z, Dong CH, Chen Y, Xiao YF, Sun FW, Guo GC (2016) Opt Lett 41(6):1249
22. Dong C, Zhang J, Fiore V, Wang H (2014) Optica 1(6):425
23. Dong C, Fiore V, Kuzyk MC, Wang H (2012) Science 338(6114):1609
24. Liu Y, Davanco M, Aksyuk V, Srinivasan K (2013) Phys Rev Lett 110(22):223603
25. Li M, Pernice WH, Tang HX (2009) Phys Rev Lett 103(22):223901
26. Dong C, Wang Y, Wang H (2015) Natl Sci Rev 2(4):510
27. Bagheri M, Poot M, Li M, Pernice WP, Tang HX (2011) Nat Nanotechnol 6(11):726

# Chapter 5
# Brillouin-Scattering-Induced Transparency and Non-reciprocal Light Storage

**Abstract** In the previous experiment, we focus on the radiation-pressure-driven mechanical oscillator. Here, we study the optomechanical interaction between the travelling optical and mechanical whispering-gallery modes. We demonstrate the Brillouin-scattering-induced transparency in a high-quality whispering-gallery-mode optical microresonator. The triply resonant Stimulated Brillouin scattering process underlying the Brillouin-scattering-induced transparency greatly enhances the light-acoustic interaction, enabling the storage of light as a coherent, circulating acoustic wave with a lifetime up to 10 μs. Furthermore, because of the phase-matching requirement, a circulating acoustic wave can only couple to light with a given propagation direction, leading to non-reciprocal light storage and retrieval. These unique features establish a new avenue toward integrated all-optical switching with low-power consumption, optical isolators, and circulators.

## 5.1 Introduction

Stimulated Brillouin scattering (SBS) is a fundamental interaction between light and traveling acoustic waves that primarily arises from electrostriction and photoelastic effects, with an interaction strength orders of magnitude greater than other relevant nonlinear optical processes. Here, we report an experimental demonstration of Brillouin-scattering-induced transparency (BSIT) in a high-quality whispering-gallery-mode optical microresonator. The triply resonant SBS process underlying the BSIT greatly enhances the light-acoustic interaction, enabling the storage of light as a coherent, circulating acoustic wave with a lifetime of up to 10 μs. Furthermore, because of the phase-matching requirement, a circulating acoustic wave can only couple to light with a given propagation direction, leading to non-reciprocal light storage and retrieval. These unique features establish new avenues toward integrated all optical switching with low-power consumption, optical isolators and circulators.

Stimulated Brillouin Scattering (SBS) in fibers and waveguides has attracted strong interest for a variety of photonic applications in past decades [1, 2], such as light storage [3], slow light [4, 5], laser [6], optical isolator [7, 8]. SBS can be incorporated into integrated photonic circuits, where the tight confinement of the optical

 
Z. Shen, *Experimental Research of Cavity Optomechanics*, Springer Theses,
https://doi.org/10.1007/978-981-33-4458-7_5

fields greatly enhances the SBS interaction. Giant enhancement of SBS in the sub-wavelength limit due to radiation pressures or boundary-induced nonlinearities has been demonstrated recently [9, 10]. SBS has also been realized in optical microresonators, such as silica microspheres [11] and disks [12], crystalline cylinders [13]. In these whispering gallery resonators, optical and acoustic waves circulate along the equatorial surface, forming optical and mechanical whispering-gallery modes (WGMs) with ultrahigh quality factor. The SBS process can become triply resonant when the control field, the Stokes or anti-Stokes field, and the acoustic wave are all resonant with the relevant optical and mechanical modes. The triply resonant SBS process, along with the ultrahigh Q-factors and the small mode volume, provides new opportunities for exploring coherent light-acoustic interactions. Over the past few years, low threshold Brillouin lasers [13], Brillouin optomechanics [14, 15], and Brillouin cooling [16, 17] have been reported in such triply-resonant WGMs.

Here, we report the experimental demonstration of Brillouin-scattering-induced transparency (BSIT) and non-reciprocal light storage in a silica microsphere resonator. In contrast to the optomechanically induced transparency (OMIT) that has been observed in a variety of optomechanical systems [18, 19], two optical modes are on resonance with the forward Brillouin acoustic mode in BSIT. A strong optical control field on the lower frequency optical WGM produced a coherent interaction between the acoustic and the other optical WGMs, which induced a transparency window for the probe light. A number of remarkable coherent quantum phenomena and potential applications are possible, such as light storage, dark modes and frequency conversion [20–22]. Due to the requirement of phase match for traveling waves in SBS, the coherent photon-phonon conversion is only allowed for waves propagating in certain direction. Benefit from this, the retrieval and non-reciprocal light storage are demonstrated. These results reveal that SBS is an excellent candidate for components in photonic integrated circuits. In particular, the acoustic vibration can be cooled to the ground state for the applications at the single photon level, such as single photon storage and frequency conversion [23–25].

## 5.2 Stimulated Brillouin Scattering

In a silica microsphere resonator, there are optical and acoustic WGMs that propagate along the surface, as shown in Fig. 5.1a. Both optical and acoustic WGMs are quantized by the orbital angular momentum $m$. When the acoustic WGM ($a$) and two optical WGMs ($c$ and $d$) satisfy energy and momentum conservations, such that $\omega_a = \omega_d - \omega_c$ and $m_a = m_d - m_c$, photons can be scattered between the optical resonances through Brillouin scattering. In this work, we focus on the forward SBS, in which $m_c$ and $m_d$ have the same sign and both optical modes are coupled to traveling waves in the same direction through the tapered fiber, as depicted in Fig. 5.1a. As schematically illustrated in Fig. 5.1b, the SBS of the control laser pumping on the lower-frequency optical mode leads to phonon absorption and anti-Stokes photon generation, whereas the Stokes process is inhibited. By contrast, the probe light

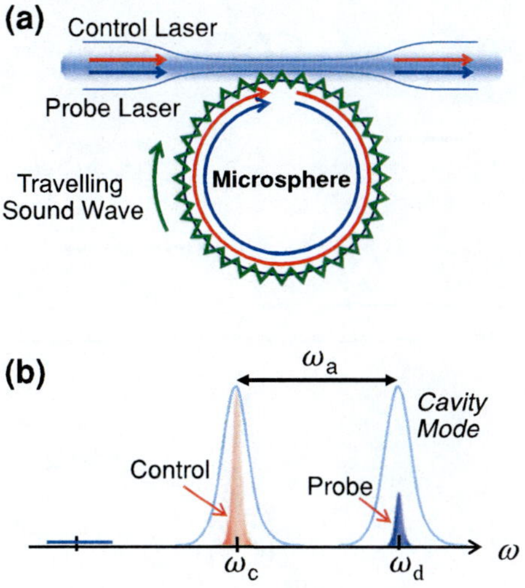


**Fig. 5.1** **a** Schematic illustration of the light-acoustic wave interaction in a microsphere. The optical modes of the microsphere are excited by the control and probe lasers through the tapered fiber, and these modes interact with the traveling acoustic wave due to the forward Brillouin scattering. **b** Spectrum diagram of the coherent photon-phonon interaction: the control laser is near resonance with the lower-frequency mode, and the probe light is on resonance with another cavity mode, whereas the Stokes process is suppressed

near-resonant with higher frequency optical mode generates phonons and Stokes photons.

The triply resonant SBS was found for the laser wavelength being approximately 1562 nm in a silica microsphere with a radius of 98 μm. The scattered anti-Stokes light was detected by measuring its beating signal with the control laser. The corresponding spectral line was monitored using an electrical spectrum analyzer, as shown in Fig. 5.2b. The Lorentz-shaped peak indicates an acoustic WGM with a frequency of $\omega_a/2\pi = 42.3$ MHz and a linewidth of $\gamma_a/2\pi = 4$ kHz, corresponding acoustic quality factor $Q_a \approx 10600$.

We used a controlled optical pumping wavelength of 1562 nm for a silica microsphere cavity with a radius of approximately 98 μm, and observed a three-mode resonance stimulated Brillouin scattering process. By detecting the beat frequency signal between the control light and the reverse Stokes, the generation of the reverse Stokes light can be determined. The power spectrum of the beat frequency signal when Brillouin scattering occurs can be observed with an electronic spectrum analyzer. Figure 5.2b is the measurement result. The Lorentz line in the power spectrum shows that the acoustic mode frequency is $\omega_a/2\pi = 42.3$ MHz and the line width is $\gamma_a/2\pi = 4$ kHz, corresponding mechanical vibration mode quality factor $Q_a \approx 10600$. The acoustic WGM was also further verified by measuring the Stokes scattering. From the acoustic velocity of silica, the orbital angular momentum of the acoustic WGM was determined to be $m_a = 6$, as shown in the inset of Fig. 5.2b.

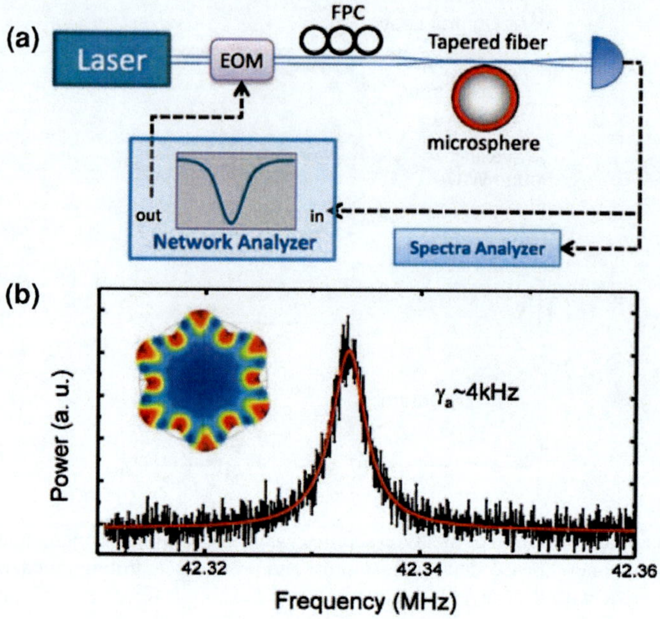


**Fig. 5.2** **a** The experimental setup. EOM, electro-optical modulator; FPC, fiber polarization controller. **b** The spectrum of one typical acoustic mode at 42.3 MHz in the microsphere when only the control laser was fixed on resonance with $\omega_c$. Inset: simulated deformation of microsphere at the equator for acoustic WGM with $m_a = 6$

All experiments were performed at room temperature and atmospheric pressure. Silica microspheres were fabricated by melting a tapered fiber with a $CO_2$ laser. Optical WGMs in the microsphere were excited through the evanescent field of a tapered optical fiber with a tunable narrow linewidth external cavity laser at the 1550 nm band. The coupling strength between the tapered fiber and the optical WGMs could be adjusted by changing the air gap between them, which was controlled by a high-resolution translation stage. The output light was detected using a low-noise photoreceiver, which was connected to a digital oscilloscope to measure the transmission spectra or to a spectra analyzer to find the acoustic modes.

When the two optical modes ($c$ and $d$) and the acoustic mode satisfy the triply resonant condition, there are actually two different processes of the Brillouin scattering:

(1) The control laser is coupled to lower-frequency mode $c$: the anti-Stokes scattered light is coupled to mode $d$ and accompanied with absorption of acoustic wave. This process cools the amplitude of acoustic wave.

(2) The control laser is coupled to higher-frequency mode $d$: the Stokes scattering process leads to the generation of phonon-photon pair, corresponding to the parametric down conversion. This process amplifies (heats) the amplitude of acoustic wave.

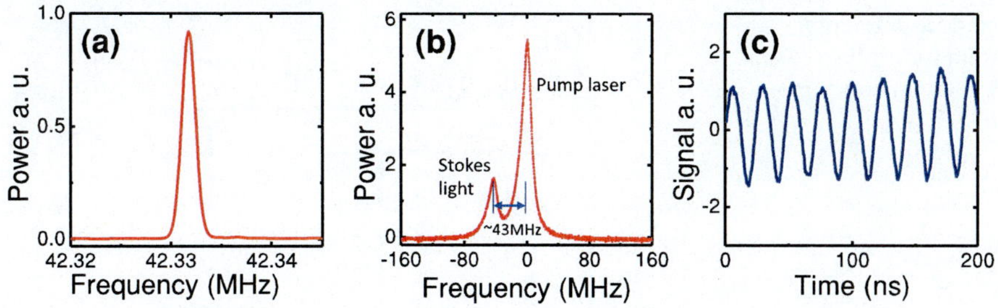


**Fig. 5.3** Optical and electrical spectra for Brillouin scattering by a 42.3 MHz acoustic wave. **a** The optical spectrum measured by Fabry–Perot spectrum analyzer. **b** The time-dependent detected optical power represents the beating signal between the two optical signals. **c** The corresponding spectrum of the results in **b**. These results are obtained in a sphere of diameter 196 $\mu$m with a control wavelength around 1562 nm

For the first case, Fig. 5.2b shows the spectra obtained from spectrum analyzer for the control laser fixed on-resonance with mode $c$. The peak was originated from the beating between control and scattered anti-Stokes lights. Due to low thermal excitation of acoustic phonon, the anti-Stokes signal is very weak compared with the control field.

To confirm the observation of the Brillouin scattering, we also studied the second case. Since the acoustic wave is amplified, the Stokes signal is much stronger than the first case. To demonstrate the Brillouin scattering induced Stokes lights directly, we measured the optical spectrum by a tunable Fabry–Perot spectrum analyzer (Thorlabs SA200-14A). Shown in Fig. 5.3a is the spectrum of cavity emission for strong control laser driving on-resonance with mode $d$. There is a prominent peak on the red side of the control light, indicating the Stokes scattering light. The frequency difference between the peaks is about 43 MHz, which agrees well with the results by beating signal as shown by Fig. 5.3b, c. Additionally, the absence of the anti-Stokes sideband in the optical spectrum as well as the absence of higher harmonics clearly confirms the triply resonant condition.

## 5.3 Brillouin-Scattering-Induced Transparency

In the demonstrated triply-resonant SBS above, a variety of coherent quantum phenomena, such as Brillouin scattering induced transparency (BSIT), in analogue to the electromagnetically induced transparency in atomic system, can be expected due to the resonant enhancement. The mechanism of BSIT can be understood physically as follows. The presence of control light and probe light in the system excites coherent vibrations of the acoustic mode, leading to scatter the control light to produce anti-Stokes light. The phenomenon of BSIT is caused by the interference between anti-Stokes light and probe light. The system Hamiltonian including the nonlinear photon-phonon interaction can be written as

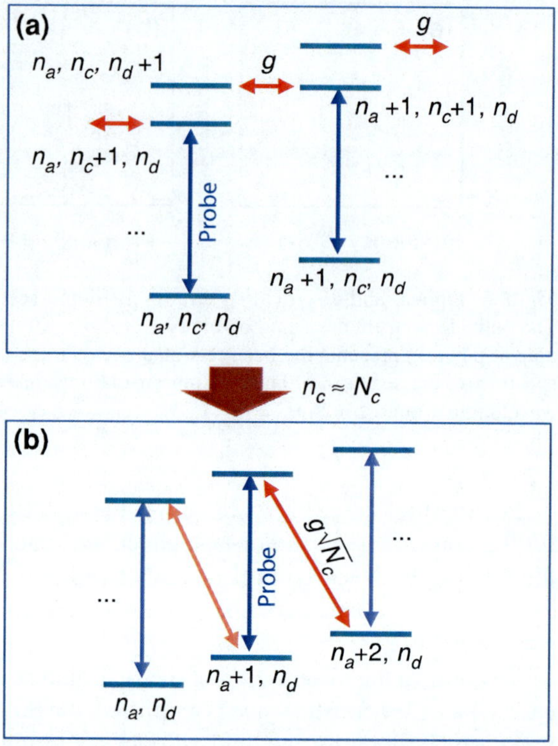


**Fig. 5.4** **a–b** Energy diagram for triply-resonant SBS in the microsphere

$$H = \omega_a a^\dagger a + \omega_c c^\dagger c + \omega_d d^\dagger d + g(a^\dagger c^\dagger d + acd^\dagger), \tag{5.1}$$

where $a$ is Boson operator of the acoustic mode, and $c$, $d$ are Boson operators of the optical modes, $g$ is the vacuum SBS strength which is nonzero only when the phase matching condition is satisfied. The energy diagram of the system is schematically illustrated in Fig. 5.4a, where the energy levels are described by phonon and photon Fock states $|n_a, n_c, n_d\rangle$, where $n_{a(c,d)}$ is the number of phonons (photons). SBS induces transitions between $|n_a, n_c, n_d + 1\rangle$ and $|n_a + 1, n_c + 1, n_d\rangle$, thereby leading to Boson annihilation and the creation of all three modes. Due to the Boson nature of photon, the transitions is enhanced by a factor proportional to the control photon number. For the strong control field that $\sqrt{n_c} \approx \sqrt{n_c + 1} \approx \sqrt{N_c}$, the energy diagram can be simplified as in Fig. 5.4b. The model incorporates the well-known electromagnetically induced transparency in a $\Lambda$-type atom: the probe light induces the transition of photon mode $d$, while the energy level is strongly coupled to other states through phonon transition. Therefore, the rising destructive interference of excition pathways for the intracavity field leads a transparency window of the probe field, corresponding to the BSIT.

Under the strong control laser driving, the Hamiltonian is effectively simplified to:

$$H = -\delta a^{\dagger}a - (\delta + \Delta)d^{\dagger}d + g\sqrt{N_c}(a^{\dagger}d + ad^{\dagger}) + i\sqrt{\kappa_{d,1}}\epsilon_p(d^{\dagger} - d), \tag{5.2}$$

Here, $\epsilon_p$ is a weak probe light with a frequency of $\omega_p$ coupling to mode $d$, $\Delta = \omega_a + \omega_l - \omega_d$ and $\delta = \omega_p - \omega_l - \omega_a$ are frequency detunings, where $\omega_l$ is the frequency of the control laser. The steady-state intracavity power spectrum can be solved as follows:

$$I_d(\delta) \propto \left| \frac{\kappa_{d,1}}{i(\delta + \Delta) - \kappa_d/2 + \frac{g^2 N_c}{i\delta - \gamma_a/2}} \right|^2 . \tag{5.3}$$

The intracavity power of the optical mode $d$ is modified by the coherent photon-phonon interaction, giving rise to changes in the intracavity field when $\frac{g^2 N_c}{i\delta - \gamma_a/2}$ is comparable to $\kappa_d$. It is convenient to introduce the cooperativity factor $C = \frac{4g^2 N_c}{\gamma_a \kappa_d}$ to evaluate the coherent interaction strength.

To verify the BSIT in our system, we probed the intracavity power spectrum using the heterodyne detection, with the control beam serving as the local oscillator. The probe light was generated from the modulated control laser using an EOM, which was modulated at a frequency of $\Omega$. The demonstration of the total heterodyne signal with the modulation frequency $\Omega$ using a network analyzer allowed the intracavity power to be extracted from the heterodyne signal, which is related to the probe field through a simple relation. As shown in Fig. 5.5a–c, we investigated the dependence of the transparency window on the detuning $\Delta$ by adjusting $\omega_l$ with a fixed pumping power of 0.3 mW. The shift of the transparency window obeyed the triple-resonance condition, and the asymmetric Fano-type lineshapes were in good agreement with theoretical model (Eq. (5.3)). We estimated values of $C = 5.6,\ 0.71,\ 0.26$ for $\Delta = 1.3,\ 0,\ -1.3$ MHz, respectively.

The strongest coupling rate was not at a detuning of $\Delta = 0$ because the triply-resonant condition was not exactly satisfied, as estimated $\omega_d - \omega_c - \omega_a \approx 1.3$ MHz. Further studies of the BSIT dip at $\Delta = 0$ are presented in Fig. 5.6a, b for control laser powers ranging from 180 to 350 $\mu$W. Dips of increasing depths and widths were observed, which could be fitted with a simple Lorentz function. According to Eq. (5.3), the expected linewidth linearly increases with $C$ proportional to $(1 + C)\gamma_a$, and the minimum heterodyne signal of the BSIT is given by $\frac{1}{(1+C)^2}$. However, we found the $C$ was not proportional to the input power because of the thermal effect, which changed the triply-resonant condition. The triply resonant Brillouin Scattering system enables us to achieve a cooperativity near 6 with a control laser power about 300 $\mu$W. Much greater control powers are needed for the achievement of similar cooperativity in WGM-type optomechanical resonators that do not satisfy the triple resonant conditions.

In our experimental system, the system consist of: (1) acoustic wave $a$, (2) two optical cavity modes $c$ and $d$. There are two lasers input to the system, the control laser near resonant to cavity mode $c$ and the probe laser near resonant to cavity

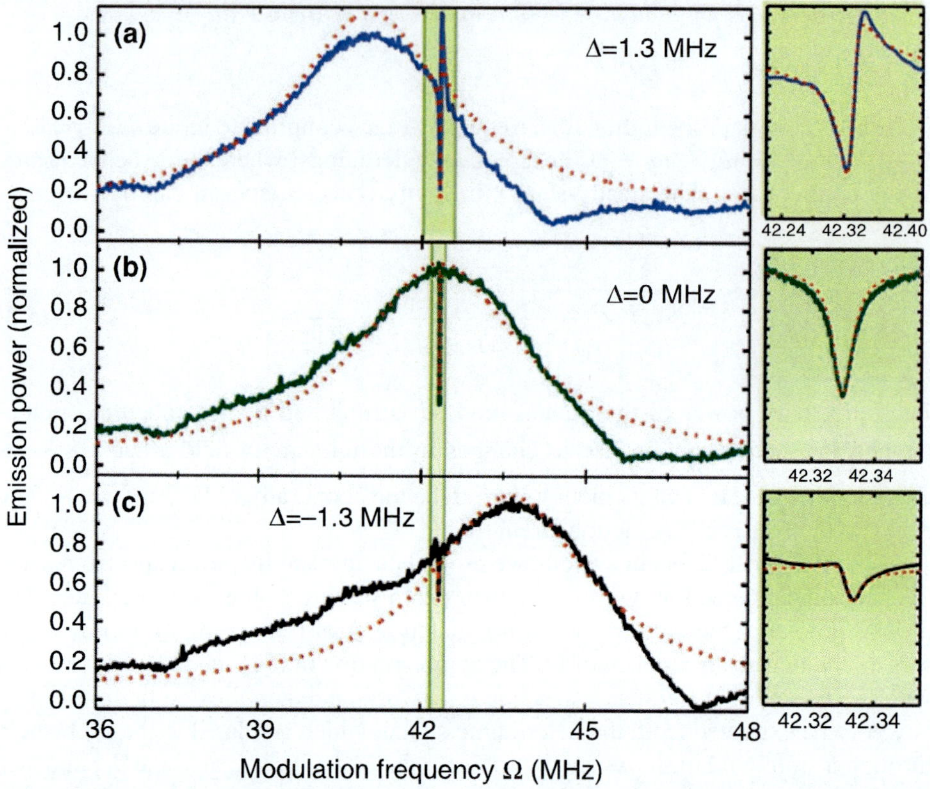


**Fig. 5.5** **a–c** Experimentally observed, normalized heterodyne traces acquired when the probe frequency was scanned by sweeping the phase modulator frequency for several values of control-beam detuning, $\Delta = 1.3,\ 0,\ -1.3$ MHz. Although the center of the response of the bare optical cavity shifted correspondingly, the characteristic sharp dip of BSIT was always observed for $\delta = 0$. The input power of the control beam launched into the cavity during the measurements was 0.3 mW. The right column shows the spectral response in the relevant anti-Stokes resonance with an expanded frequency scale. The short-dashed lines are the results of calculation results

mode $d$. The whole system is inevitably complicated, thus we plotted the detailed illustration of the modes and lasers in Fig. 5.7.

The two cavity modes are shown by Red and Blue Lorentz peaks. The control laser is denoted by a narrow pink peak, with the frequency of $\omega_l$. The probe laser is generated by EOM (modulation frequency $\Omega$) sideband of the control laser, as shown by the blue tiny peaks around the control laser.

The triple resonance is satisfied when $\omega_d - \omega_c = \omega_a$, corresponding to $\Delta_c$ and $\Delta$ both equal to 0.

The Brillouin scattering induced transparency happens when $\omega_p - \omega_l = \omega_a$, corresponding to $\delta = 0$.

Below we present a detailed theoretical derivation of the Brillouin scattering-induced transparency process:

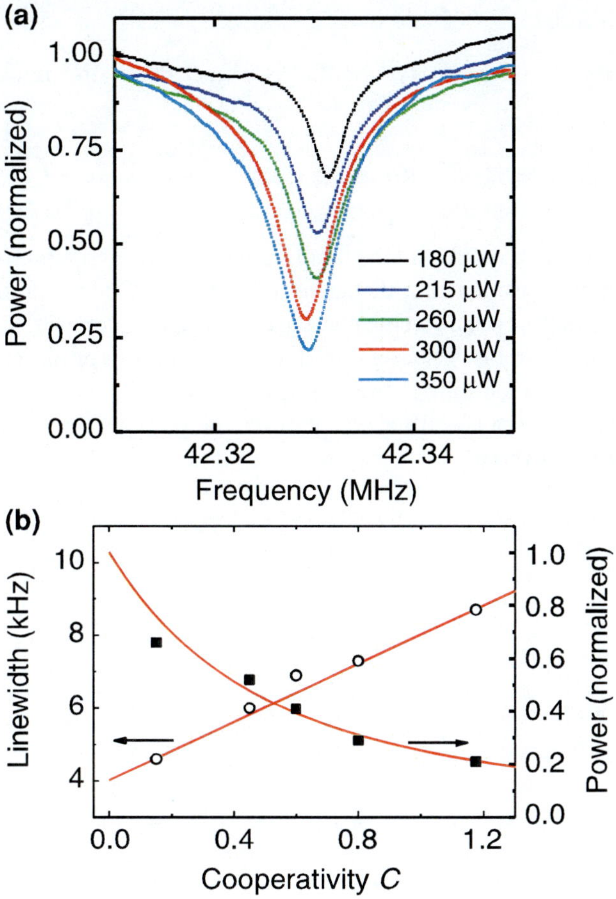


**Fig. 5.6** **a** The emission power from the WGM near the anti-Stokes resonance for five different powers of the control beam from 0.18 mW up to 0.35 mW. **b** The spectral linewidth of the BSIT dip and the power emitted by the WGM as functions of the mechanical cooperativity $C$, as derived from **b**. The solid lines in **b** represent the theoretically expected values

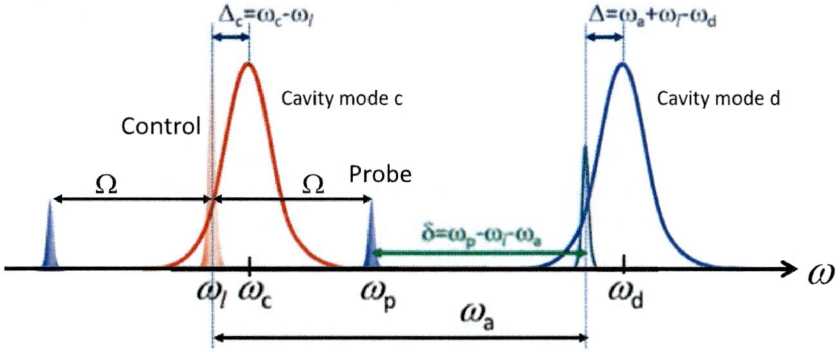


**Fig. 5.7** Optical frequency relationship of the coupled triple system for Brillouin-scattering-induced transparency experiment

Optical Cavity Modes

For the microsphere microcavity coupling to taper fiber, the interaction can be described by

$$H = \omega_c c^\dagger c + \omega_d d^\dagger d + i\sqrt{\kappa_{c,1}}\epsilon_l (c^\dagger e^{-i\omega_l t} - c e^{i\omega_l t}) + i\sqrt{\kappa_{d,1}}\epsilon_p (d^\dagger e^{-i\omega_p t} - d e^{i\omega_p t}). \tag{5.4}$$

The operators $c$ ($c^\dagger$) and $d$ ($d^\dagger$) are the annihilation (creation) operators of the optical cavity. The parameter $\kappa_{c,1}$ and $\kappa_{d,1}$ is the coupling strength between the cavity mode and waveguide, which depends on the overlap of electromagnetic field between cavity and waveguide mode fields. Usually, the spatial distribution of optical modes $c$ and $d$ are different, thus the coupling strengths are different.

For the system without Brillouin scattering interaction, the control and probe cavity field can be solved as

$$\frac{d}{dt}c = (-i\omega_c - \kappa_c/2)c + \sqrt{\kappa_{c,1}}\epsilon_l e^{-i\omega_l t}, \tag{5.5}$$

$$\frac{d}{dt}d = (-i\omega_d - \kappa_d/2)d + \sqrt{\kappa_{d,1}}\epsilon_l e^{-i\omega_p t}. \tag{5.6}$$

The transmissions of the laser are

$$E^c_{out} = \epsilon_l - \sqrt{\kappa_{c,1}}c, \tag{5.7}$$

$$E^d_{out} = \epsilon_p - \sqrt{\kappa_{d,1}}d. \tag{5.8}$$

Linearized Hamiltonian

When the phase matching condition is satisfied, the forward SBS leads to coherent interaction between the optical and acoustic modes. The Hamiltonian of the system with control and probe light input is

$$\begin{aligned} H = {} & \omega_a a^\dagger a + \omega_c c^\dagger c + \omega_d d^\dagger d + g(a^\dagger c^\dagger d + a c d^\dagger) \\ & + i\sqrt{\kappa_{c,1}}\epsilon_l (c^\dagger e^{-i\omega_l t} - c e^{i\omega_l t}) + i\sqrt{\kappa_{d,1}}\epsilon_p (d^\dagger e^{-i\omega_p t} - d e^{i\omega_p t}). \end{aligned} \tag{5.9}$$

Here, the $a$ and $a^\dagger$ denote the operators of acoustic wave.

For very strong control laser pumping to mode c, under the condition that the probe laser power and acoustic wave amplitude are very weak, we can treat the mode field classically as complex number. Under the non-depletion approximation of the nonlinear three-wave mixing, the dynamics of control mode c can be described as there is no influence by SBS, as the same as Eq. (5.5). In our experiment, the duration of control pulse $\tau_{pump} \gg 1/\kappa_c$, thus we have the steady state solution of mode $c$ as

$$\langle c \rangle = \sqrt{N_c} e^{-i\omega_l t} = \left| \frac{\sqrt{\kappa_{c,1}}\epsilon_l}{-i\Delta_c - \kappa_c/2} \right| e^{-i\omega_l t}, \tag{5.10}$$

where the detuning $\Delta_c = \omega_c - \omega_l$.

By substituting this complex number into the Hamiltonian, we obtain the linearized photon-phonon interaction as

$$H = \omega_a a^\dagger a + \omega_d d^\dagger d + g\sqrt{N_c}(ad^\dagger e^{-i\omega_l t} + a^\dagger d e^{i\omega_l t}) + i\sqrt{\kappa_{d,1}}\epsilon_p(d^\dagger e^{-i\omega_p t} - de^{i\omega_p t}). \tag{5.11}$$

In the rotating frame of $H_0 = \omega_p d^\dagger d + (\omega_p - \omega_l)a^\dagger a$, we can get:

$$H_{eff} = -\delta a^\dagger a - (\delta + \Delta)d^\dagger d + g\sqrt{N_c}(a^\dagger d + ad^\dagger) + i\sqrt{\kappa_{d,1}}\epsilon_p(d^\dagger - d), \tag{5.12}$$

where the detuning are $\delta = \omega_p - \omega_l - \omega_a$, $\Delta = \omega_a + \omega_l - \omega_d$.
Brillouin Scattering Induced Transparency
Under the strong control laser field, the optical mode d and acoustic mode a are coupled linearly. Therefore, coherent conversion between the photon and phonon can be realized. The dynamics of the probe light mode and acoustic mode are:

$$\frac{d}{dt}a = (i\delta - \gamma_a/2)a - ig\sqrt{N_c}d, \tag{5.13}$$

$$\frac{d}{dt}d = [i(\delta + \Delta) - \kappa_d/2]d - ig\sqrt{N_c}a + \sqrt{\kappa_{d,1}}\epsilon_p. \tag{5.14}$$

In the steady state $\frac{d}{dt}a = \frac{d}{dt}d = 0$, we have

$$a = \frac{ig\sqrt{N_c}}{i\delta - \gamma_a/2}d, \tag{5.15}$$

$$d = \frac{-\sqrt{\kappa_{d,1}}\epsilon_p}{i(\delta + \Delta) - \kappa_d/2 + \frac{g^2 N_c}{i\delta - \gamma_a/2}}. \tag{5.16}$$

From the experiments, the measured heterodyne signal is corresponding to the intensity of intracavity field:

$$I_d(\delta) \propto \left| \frac{\kappa_{d,1}}{i(\delta + \Delta) - \kappa_d/2 + \frac{g^2 N_c}{i\delta - \gamma_a/2}} \right|^2. \tag{5.17}$$

## 5.4 Non-reciprocal Light Storage

Owing to the coherent SBS interaction, the coherent conversion between photons and acoustic phonons can be used for light storage. In contrast with the previously studied optomechanics driven by radiation and gradient forces, in which mechanical

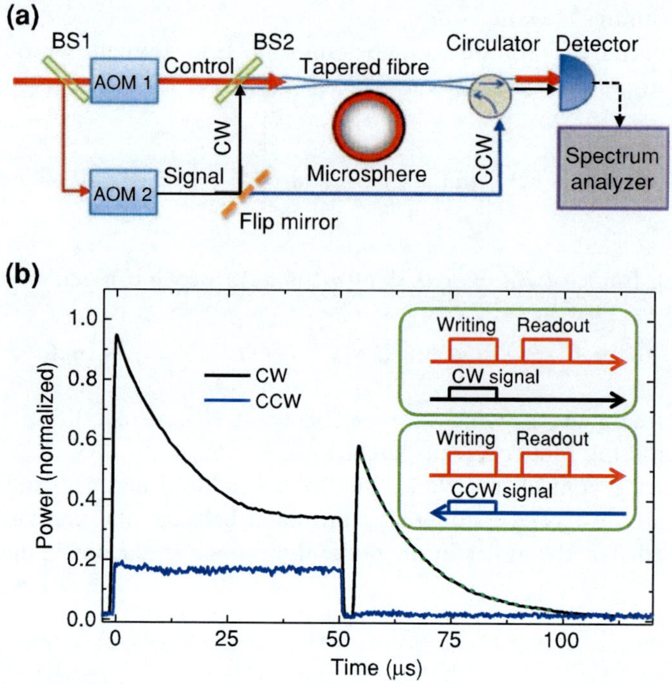


**Fig. 5.8** **a** The experimental setup of the non-reciprocal light storage. The CW signal is combined into the fiber through the beam splitter (BS2). The CCW signal is launched into the fiber through the circulator. **b** The measured intracavity signal light during the storage and retrieval processes. The black and blue lines correspond to the input signal from different input directions. Inset: the pulse sequences for writing, readout and signal, the arrows indicate the propagation direction

vibrations can couple to a variety of optical modes, Brillouin scattering is possible only for specific modes that satisfy certain energy and momentum conservation conditions. The input control light coupling to clockwise (CW) optical mode only permits interaction between CW acoustic phonons and photons, whereas the counter-clockwise (CCW) probe light or acoustic wave is not affected by SBS. Therefore, the requirement of phase matching for nonlinear SBS process gives rise to the breaking of time reversal symmetry of light, which is switchable by control laser.

To verify this non-reciprocity, we studied the light storage of CW or CCW signal photons with a fixed CW control light. As shown in Fig. 5.8a, the control laser (through AOM 1) is coupled to CW mode only, and the signal light (through AOM 2) is coupled to CW or CCW mode by moving the flip mirror. When the triply-resonant condition is satisfied, the CW signal photons will be converted to CW acoustic phonons coherently, as indicated by the beam-splitter type Hamiltonian. As is evident in Fig. 5.8b, the CW signal (black line) was stored during the writing pulse and retrieved after 3 μs. During the writing pulse, the intracavity signal photons show an exponential decay, which is due to interference between the photon and exponentially increasing phonons, and a similar phenomenon has been reported for

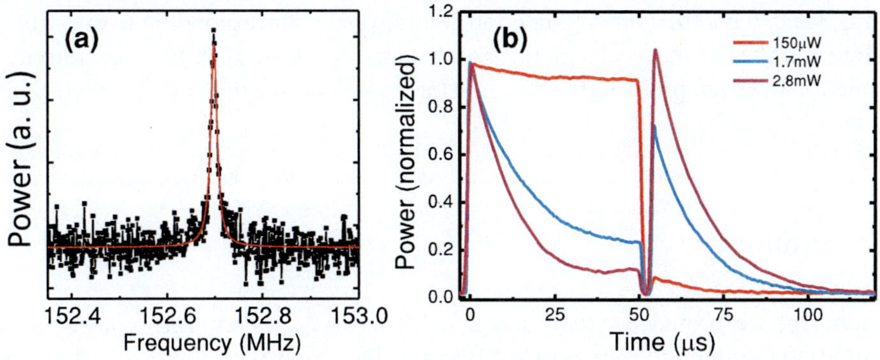


**Fig. 5.9** **a** The spectra of the acoustic mode used for the light storage experiment. The resonance frequency is $\omega_a/2\pi = 152.7$ MHz and the linewidth is $\gamma_a/2\pi = 15$ kHz, corresponding to the acoustic quality factor $Q_a = 10180$. **b** The storage and retrieval of emission power with various pump power

the breathing mode. After 3 μs storage, the phonons are converted to signal photons due to the readout pulse. The decay time of converted signal was calculated with 14 μs, as shown by the green dashed.

For the CCW launched signal (blue line), there was no light retrieval during the readout pulse. It should be noted that the flat signal in the writing pulse originated from the beating between the writing photons and the slight reflected photons of the signal at the end face of the circulator.

The microsphere sample is changed in the experiment for the light storage. The acoustic mode that satisfy the energy and momentum conservations is measured by the spectra analyzer with strong control laser, with the similar procedure as the BSIT experiments. Shown in Fig. 5.9a is the spectrum, indicating a high-Q acoustic mode with frequency of $\omega_a/2\pi = 152.7$ MHz and the linewidth of $\gamma_a/2\pi = 15$ kHz, corresponding to an acoustic Q-factor $Q_a = 10180$. As shown in Fig. 5.8a, The control and probe light are generated by two separated acousto-optic modulators (AOM 1 and 2). The writing and readout control pulse sequence was generated by AOM 1, in which a laser beam was frequency-shifted by −80 MHz. The signal pulse was obtained by AOM 2 with frequency-shifted by +72.7 MHz, which was in synchronization with the writing pulse. The peak powers of the writing and readout pulses were each 1 mW and the peak power of signal pulse was 20 nW. The durations of writing and reading pulse were 50 and 80 μs. The two pulses were separated by 3 μs as a storage time. The forward and backward signal pulses were separately coupled in the fiber. The power of the emitted light was measured using the spectrum analyzer with gate detection mode, and the resolution bandwidth (RBW) was 10 MHz. respectively.

In the above optical signal storage experiments, the results shows the signal retrial after 3 μs storage in the microsphere. More experimental data are provided in Fig. 5.9b to support the non-reciprocal light storage. Shown by different curves in Fig. 5.9b are the light storage and retrieval for various control laser power. From the results, we can find that the higher control laser power corresponding to the stronger decline and greater retrieval signal. These can be explained that the cooperativity of interaction between probe light and acoustic wave is proportional to the control laser power.

## 5.5 Summary

To sum up, we provided experimental evidences of coherent and non-reciprocal photon-phonon inter-conversion for SBS in a silica microsphere, though the experiments at room temperature are affected by the thermal acoustic excitations. Since the interaction photons and phonons with strong control field is in the form as $g\sqrt{N_c}(a^\dagger d + ad^\dagger)$, the coherent conversion between photons and phonons persists even when the signal field is at the single photon level. In particular, the single photon storage becomes possible when the relevant acoustic modes are cooled down to their ground states. For $\omega_a/2\pi = 1$ GHz and at T $= 0.5$ K, the thermal phonon occupation is near 10. In this case, the ground state cooling is achievable for silica microspheres. Actually, a close related scheme of the quantum state transfer mediated by a mechanical oscillator had been already demonstrated at cryogenic temperature recently [23]. Therefore, the demonstrated BSIT and nonreciprocity are potential for photonic devices at single photon level.

Because of the momentum conservation required for SBS, the degenerate CW and CCW acoustic modes can be written and read out independently by the corresponding CW and CCW control light, respectively. In the whispering gallery microresonators, CW and CCW optical and acoustic WGMs are degenerated. Therefore, a pair of WGMs can naturally be treated as one bit of binary information carriers. Here, we also proposed a design of quantum memory for polarization-encoded photon state. As illustrated in Fig. 5.10, with the use of PBS and PR, horizontally and vertically polarized input photon states can individually couple to CW and CCW

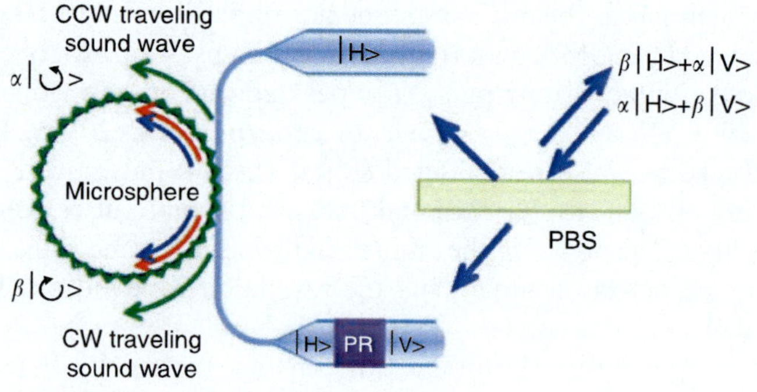


**Fig. 5.10** Schematic diagram of single-bit quantum memory based on circulating acoustic phonons. PBS: polarization beam splitter; PR: polarization rotator

optical modes. These states can be stored as the CW and CCW acoustic waves and read out, separately. In principle, the input state, a superposition of polarization state $\alpha \left|H\right\rangle + \beta \left|V\right\rangle$, can be stored as a superposition of circulating acoustic state. In the reversal of the storage process, the states can be read out by being converted to the photon state $\alpha \left|V\right\rangle + \beta \left|H\right\rangle$ after tens of microseconds.

Brillouin scattering in the whispering-gallery microresonators is promising in several aspects. (a) Brillouin scattering enables the optical coupling to acoustic waves with a frequency range from a few MHz to 11 GHz, thereby providing a diverse platform for coherent light-matter interactions. (b) The triply-resonant configure can enhance the SBS, thereby reducing the power consumptions [14, 16]. (c) The traveling wave properties provide the light non-reciprocity, thus offering potential application in all optical integrated isolator and circulator devices [26, 27]. Our studies pave the way toward the coherence coupling between photons and acoustic phonons and encourage further investigations of non-reciprocity and memory at the quantum level.

## References

1. Kobyakov A, Sauer M, Chowdhury D (2010) Adv Opt Photonics 2(1):1
2. Eggleton BJ, Poulton CG, Pant R (2013) Adv Opt Photonics 5(4):536
3. Zhu Z, Gauthier DJ, Boyd RW (2007) Science 318(5857):1748
4. Okawachi Y, Bigelow MS, Sharping JE, Zhu Z, Schweinsberg A, Gauthier DJ, Boyd RW, Gaeta AL (2005) Phys Rev Lett 94(15):153902
5. Pant R, Byrnes A, Poulton CG, Li E, Choi DY, Madden S, Luther-Davies B, Eggleton BJ (2012) Opt Lett 37(5):969
6. Abedin KS, Westbrook PS, Nicholson JW, Porque J, Kremp T, Liu X (2012) Opt Lett 37(4):605
7. Huang X, Fan S (2011) J Lightwave Technol 29(15):2267
8. Poulton CG, Pant R, Byrnes A, Fan S, Steel M, Eggleton BJ (2012) Opt Express 20(19):21235
9. Pant R, Poulton CG, Choi DY, Mcfarlane H, Hile S, Li E, Thevenaz L, Luther-Davies B, Madden SJ, Eggleton BJ (2011) Opt Express 19(9):8285
10. Shin H, Qiu W, Jarecki R, Cox JA, Olsson RH III, Starbuck A, Wang Z, Rakich PT (2013) Nat Commun 4
11. Tomes M, Carmon T (2009) Phys Rev Lett 102(11):113601
12. Li J, Lee H, Vahala KJ (2013) Nat Commun 4
13. Grudinin IS, Matsko AB, Maleki L (2009) Phys Rev Lett 102(4):043902
14. Bahl G, Zehnpfennig J, Tomes M, Carmon T (2011) Nat Commun 2
15. Bahl G, Kim KH, Lee W, Liu J, Fan X, Carmon T (2013) Nat Commun 4
16. Bahl G, Tomes M, Marquardt F, Carmon T (2012) Nat Phys 8:203
17. Tomes M, Marquardt F, Bahl G, Carmon T (2011) Phys Rev A 84(6):063806
18. Weis S, Riviere R, Deleglise S, Gavartin E, Arcizet O, Schliesser A, Kippenberg TJ (2010) Science 330(6010):1520
19. Safavi-Naeini AH, Alegre TPM, Chan J, Eichenfield M, Winger M, Lin Q, Hill JT, Chang DE, Painter O (2011) Nature 472:69
20. Dong C, Fiore V, Kuzyk MC, Wang H (2012) Science 338(6114):1609
21. Hill JT, Safavi-Naeini AH, Chan J, Painter O (2012) Nat Commun 3:1196
22. Fiore V, Yang Y, Kuzyk MC, Barbour R, Tian L, Wang H (2011) Phys Rev Lett 107(13):133601
23. Palomaki T, Harlow J, Teufel J, Simmonds R, Lehnert K (2013) Nature 495(7440):210
24. Galland C, Sangouard N, Piro N, Gisin N, Kippenberg TJ (2014) Phys Rev Lett 112(14):143602

25. Tian L, Wang H (2010) Phys Rev A 82(5):053806
26. Yu Z, Fan S (2009) Nat Photonics 3(2):91
27. Li E, Eggleton BJ, Fang K, Fan S (2014) Nat Commun 5

# Chapter 6
# Optomechanical Microcavity Packaging

**Abstract** For the quantum optomechanics research, the mechanical oscillator needs to be cooled to ground state in a cryostat or vacuum chamber, which can also improve the $Q$ factors of the mechanical modes. However, stable 3D optical stages are required to hold the microresonator-taper coupling system to keep and adjust the air gap between them and, thus, give rise to experiment difficulties in a relatively small vacuum chamber or cryostat. To address this issue, an efficient method to package a coupled silica microsphere and tapered fiber system is proposed and demonstrated experimentally. For the purpose of optomechanical studies, high-quality-factor optical and mechanical modes are maintained after the packaging process. For the mounted microsphere, the coupling system is more stable and compact and, thus, is beneficial for future studies and applications based on optomechanical interactions. Especially, the packaged optomechanical system, which is tested in a vacuum chamber, paves the way toward quantum optomechanics research in cryostat.

## 6.1 Introduction

With the in-depth study of cavity optomechanics, we need to use vacuum or low temperature environment to improve the quality factor of mechanical mode, reducing the thermal noise, cooling the mechanical motion to the ground state, and performing quantum cavity optomechanical research. However, the material absorption limited high-Q optical WGMs can only be excited efficiently through near field couplers [1], such as silica fiber taper. Stable 3D optical stages are required to hold the microresonator-taper coupling system to keep and adjust the air gap between them, thus give rise to experiment difficulties in a relatively small vacuum chamber or cryostat [2]. To address this issue, several packaging methods have been proposed, and the high-Q WGM microresonators for practical sensing, filter applications, or microlasing are demonstrated [3–8]. Since people only concerns about the optical properties of the WGMs, the properties of the mechanical vibration of these microresonators are overlooked in previous package schemes. For example, extra polymer glue or the fiber taper adhere to the microresonator will destroy the mechanical Q significantly.



Z. Shen, *Experimental Research of Cavity Optomechanics*, Springer Theses,
https://doi.org/10.1007/978-981-33-4458-7_6

In this experiment, we present the packaged microsphere system maintaining high optical and mechanical Q factor. Our experiments demonstrated that how to keep high optical and mechanical Q-factors when the tapered fiber is contacted to the microsphere. Other advantages of this packaging systems include simple operation to package, long time stabilization in ambient, even in a vacuum chamber. In addition, the package framework is made by the same material silica, thus the whole system is insensitive to temperature changes. Therefore, we believe that our packaged system is promising for future experiment in the cryostat without 3D translation stage, avoiding the complicated imaging system

In this experiment, we propose a scheme for packaging a microsphere cavity without the need for a translation stage to adjust the position of the microsphere cavity and the fiber cone, while maintaining a very high optical and mechanical quality factor. The other advantages of this packaging solution are easy operation and simple steps. The encapsulated optomechanical microcavity has long-term stability in normal environments and in vacuum chambers. In addition, the materials used in the packaging system are silicon dioxide, so the response to temperature changes is consistent. This microcavity packaging avoids the use of three-dimensional high-precision translation stages and complex imaging systems, and lays the foundation for future optomechanical experiments at low temperatures [2].

## 6.2 Experimental Setup and Packaging

Figure 6.1a shows the schematic of the experimental setup. The silica microspheres with diameter about 50 μm were fabricated by melting the tapered tip of a fiber through a continuous wave $CO_2$ laser. Generally, the mechanical $Q$ factor increases with smaller diameter [9] and longer length of the fiber stem [10]. In the inset of Fig. 6.1a, the diameter and length of the microsphere with mechanical $Q$ factor were around 7 and 40 μm. The tapered fiber was obtained by melting a standard SMF-28 silica fiber with the hydrogen flame.

The packaging procedure consists of three steps, which are shown in Fig. 6.1b–d. The first step was to fix the acrylate buffer of the glass stem which supporting the microsphere to the center shaft of the C-shape glass plate by a UV glue. After that, the microsphere is fixed to the packaged framework. Then, a fiber taper is moved to the vicinity of the microspheres by a 3D translation stage. The relative position and distance to the microsphere are adjusted precisely. During this step, the optical spectrum is monitored simultaneously to ensure an efficient coupling between the fiber taper and optical WGMs. In the last step, the tapered fiber was transferred from the home-made stage to the glass stage by the UV glue, as shown in Fig. 6.1d. The whole system was also monitored in real-time through a CCD camera during the processes of packaging, as shown in the inset of figure. The whole system was also monitored in real-time through a CCD camera during the processes of packaging, as shown in the inset of Fig. 6.1a. And Fig. 6.1e shows one typical packaged microsphere.

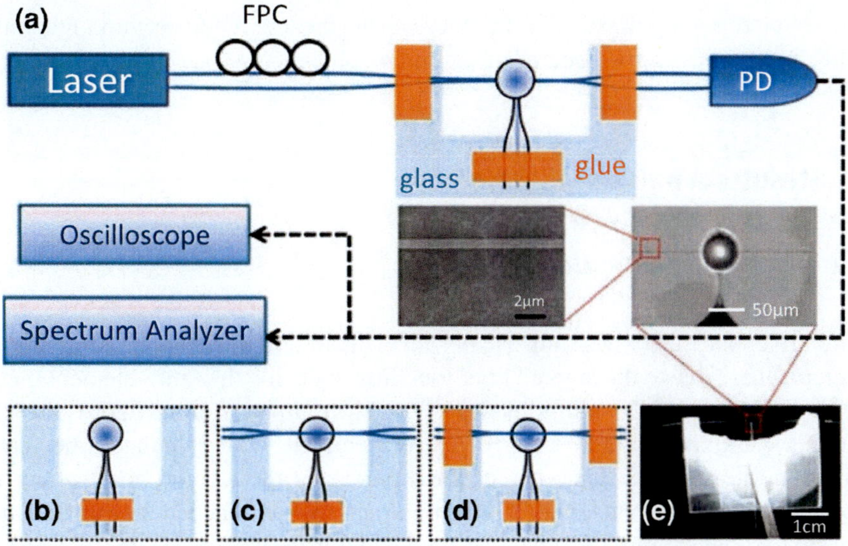


**Fig. 6.1** **a** Experimental setup for the packaged microsphere. FPC: fiber polarization controller; PD: photo detector. **b–d** Illustration of the packaging process. **e** A typical packaged microsphere-taper coupling system. Inset: SEM images of a tapered fiber and digital image of the coupled silica microsphere and tapered fiber system

Following above procedures, packaged microsphere-fiber taper system are characterized by the setup shown in Fig. 6.1a. To observe the mechanical vibrations, the tunable diode laser (1550 or 780 nm) was launched into the tapered fiber and coupled into the microspheres through evanescent field. When the probe laser is slight detuned from the cavity resonator, the thermal displacement of the mechanical resonator is transduced to the optical signal as amplitude modulated radio-frequency signals. Noting that the power of probe laser should be low enough to avoid thermal [11], Kerr [12], or optomechanical instability [13]. The output light from the tapered fiber was sent to a detector, which is connected to an electronic spectrum analyzer and a digital oscilloscope. Then, the optical transmission spectrum and the power spectrum of the mechanical vibrations are obtained. A paddle fiber polarization controller (FPC) was used to change the polarization state of input laser.

After packaging, the optical and mechanical $Q$ factor can be kept as similar as before, as discussed later. Moreover, as the packaged system is made of essentially the same glass material, the tapered fiber will remain synchronously thermal contraction as changed temperature. Thus we believe that the mounted microsphere and taper fiber system can be applied in the cryostat [14]. During the experiment, the glass plate with microsphere was risen up from underneath of the tapered fiber until mutual touching. The glass plate should be strictly parallel to the tapered fiber. Otherwise, one side of the glass plate would be first come into contact with the tapered fiber but not the microsphere. In addition, during the touching process, the microsphere could not be raised up drastically. In other words, the tapered fiber with high tension would slip away from the equator of the microsphere.

## 6.3 Results Analysis

### 6.3.1 Numeral Calculations

Before the experiments, we numerically investigated the mechanical dissipation of the microsphere due to the tapered fiber touching, by using the finite-element method (FEM) solver (COMSOL Multiphysics v4.2). The diameters of the microsphere and its stem were fixed to be 50 and 10 μm respectively. The calculated mechanical dissipation rate of the S21 modes for the diameter of the tapered fiber varied from 0.8 to 2.6 μm are shown in Fig. 6.2. If the microcavity is untouched with tapered fiber, there are two degenerated mechanical modes, with frequency around 63 MHz and mechanical dissipation rate smaller than 100 Hz. Note that in our calculation, only external radiation (clamping) losses due to tapered fiber and stems are considered, and we know the mechanical dissipation rate should be around 1.2 kHz at room temperature [15]. When tapered fiber in touching with the sphere, the dissipation rate increased significantly. An intuitive understanding is that the acoustic energy leakage by the tapered fiber proportional to the area of the fiber. Therefore, we fitted the losses that quadratically increase with the tapered fiber diameter, which are shown

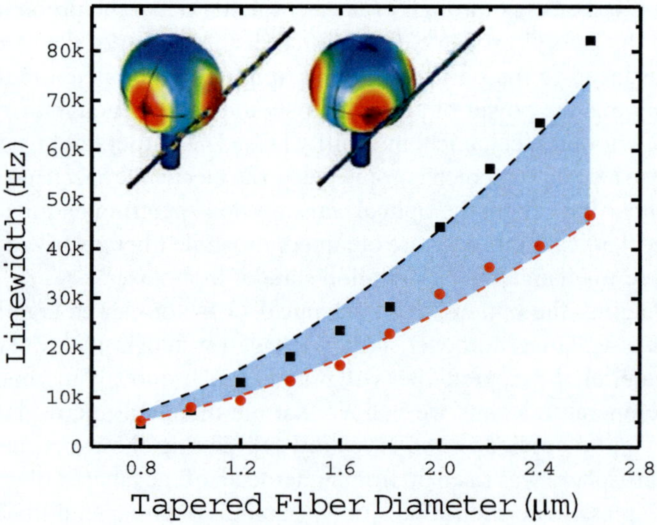


**Fig. 6.2** Numerical simulated mechanical dissipation due to fiber taper touched with the microsphere. Black and red dot are corresponding to two mechanical modes with lifted degeneracy due to taper fiber, as shown by two inset figures. Dashed lines are fitting results

by the dashed lines in Fig. 6.2. The two set of data corresponding to degenerated S21 modes, one is coupling to the tapered fiber with maximum displacement, and the other one is coupling to the tapered fiber by the node. Therefore, the mechanical dissipation rate is lower for the later mechanical mode. In practical experiments, the imperfection sphere geometry will also induce the degeneracy of S21 modes, therefore we can expect that the touched tapered fiber induced loss fall in the region between two dashed lines (the shadow area in Fig. 6.2).

### *6.3.2 Characterization by NIR Light*

To demonstrate the package and verify the effect of tapered fiber touching, we carried out the experiment at 1550 nm wavelength firstly. Figure 6.3a shows the mechanical linewidth of three different microspheres against the diameter of tapered fiber before and after the touching. The behaviour is remarkably reproducible for the different samples. Compared the linewidth to the untouched microsphere, the increment of the linewidth is about 2 kHz for taper diameter is 800 nm, which agrees well with our simulation results. For larger diameter of touching fiber tapers, we observed that the linewidth of the sample mechanical mode increases quadratically, in consist with the prediction in Fig. 6.2. Figure 6.3b, c shows typical power spectrum of the mechanical vibration of sample A with taper diameter around 800 nm. The mechanical Q factor degraded to 8770 after touching as compared with 9730 before touching. For other position, the mechanical $Q$ factor spoiled obviously close to 3 times.

The mechanisms of intrinsic mechanical dissipation of the microsphere include clamping losses, viscous damping and materials-induced losses [15, 16]. The clamping losses which are determined by the structure of the supports of the oscillator, especially the ratio between the diameter of the microsphere and its stem in our case, dominate the limit of the mechanical quality of our devices. In addition, the viscous damping which is caused by interactions with the surrounding gas atoms in our experiment, is almost constant for different spheres and can be fully eliminated in a vacuum chamber ($\rho < 1$ mbar). Therefore, due to the slightly different geometry of the spheres, the three samples have different mechanical dissipation rate and show different change of dissipation rate for the same fiber taper size.

It is worth noting that the touched optical $Q$ factors drop drastically when the coupling position is good for the touched mechanical $Q$, as shown in Fig. 6.3d, e. As we know, the optical coupling rate between the microsphere and the tapered fiber are determined by the evanescent fields overlap and phase matching conditions [17]. For the given wavelength laser, the touched optical $Q$ could be kept high at the appropriate position [3]. Otherwise, the optical $Q$ is spoiled quickly. As discussed before, the touched mechanical $Q$ could be kept with the tapered fiber diameter of 800 nm. For such size tapered fiber, the touched optical $Q$ factors could be kept with short wavelength [18]. In order to obtain high optical and mechanical $Q$ factor simultaneously, we run the experiment using the visible light.

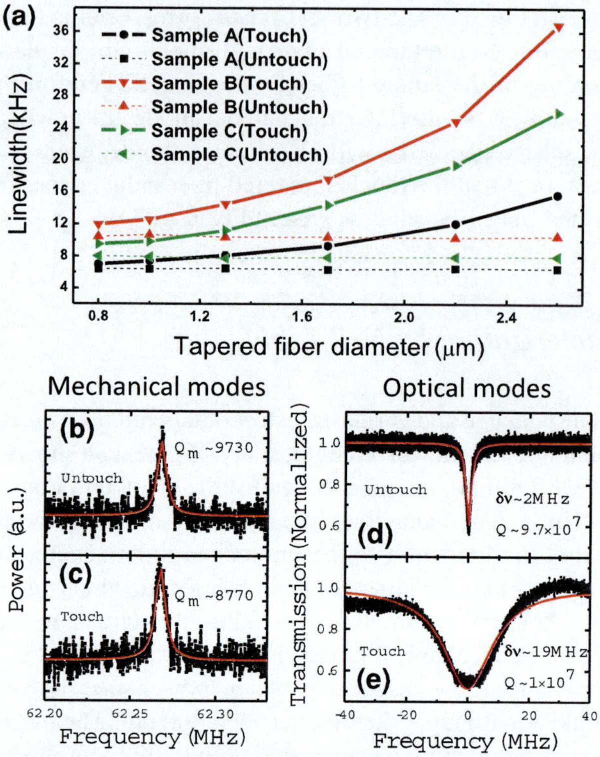


**Fig. 6.3** **a** The mechanical linewidth versus the tapered fiber diameter for three different samples before and after the touching, respectively. **b–c** The power spectrum of the mechanical vibrations before and after the touching for sample A at the position of which the tapered fiber diameter is about 800 nm. **d–e** The transmission spectrum of sample A before and after the touching at the same position as **b** and **c**. These results were all got under 1550 nm wavelength laser. The mechanical frequencies of the three samples are $\omega_m/2\pi \approx 62.2, 59.3, 57.6$ MHz

## 6.3.3 Characterization by Visible Light

In order to circumvent the dramatic decline of the optical modes, we changed the setup to 780 nm wavelength. The waist of typical 780 nm monomode tapered fiber is approximately 800 nm [19], which should spoil mechanical $Q$ slightly. Figure 6.4a shows the touched mechanical $Q$ factor for different positions on the taper while the contact point on the equator of the sphere. The interval between different positions is roughly 100 μm for the tapered fiber with a diameter of 800 nm, as shown in SEM image in Fig. 6.1a.

We also studied the touched mechanical $Q$ with different positions along the polar direction on the microsphere, as shown in Fig. 6.4b. All results indicate that the $Q$ is insensitive to the contact points. During this process, the high optical modes can be observed all the time. A typical transmission spectrum of sample A indicates the $Q$ factor up to $10^8$ (inset of Fig. 6.4a). The asymmetric lineshape in the tail of the

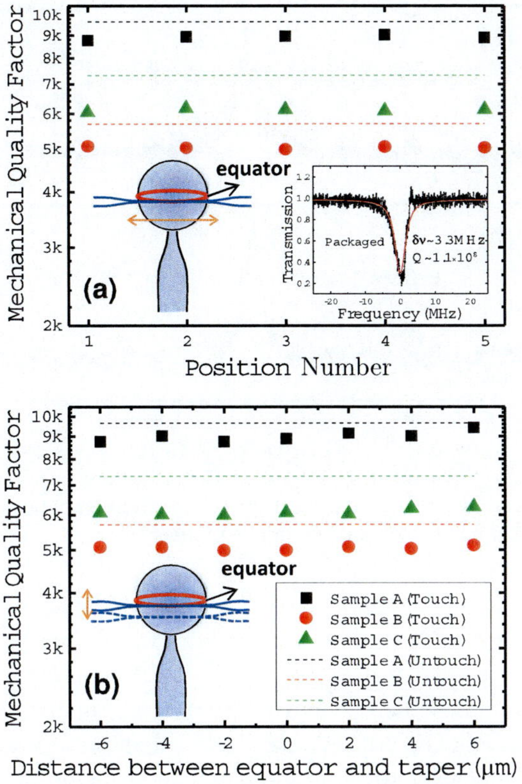


**Fig. 6.4** **a** The touched mechanical $Q$ factor of three samples at five positions of which the interval is roughly 100 μm of a typical 780 nm single mode tapered fiber, respectively. **b** The mechanical $Q$ factor of three samples versus the perpendicular distance between the equator of the microsphere and the tapered fiber. Inset: a touched transmission spectrum of sample A excited by 780 nm laser. Dashed lines indicate untouched mechanical $Q$ factor for guiding

transmission shows the ringing phenomenon [20]. The higher $Q$ factor of optical mode would pull the system in the resolved band regime and greatly decrease the threshold for the pump power ($P \propto 1/Q_0^3$), which are good for the application of cavity optomechanics. These results indicate that even without the high resolution 3D translation stage, high mechanical and optical $Q$ factor can be obtained after the packaging process. Moreover we could infer that even in the encapsulation process the micrometer order of magnitude change of the relative position, the mechanical quality would not be affected.

To verify the practicability of the mounted scheme, experiment in vacuum was also carried out. Figure 6.5a, b are the power spectrum of mechanical vibration in atmospheric pressure and in vacuum for one mounted system. It is found that the mechanical $Q$ factor is even higher than the results in air before package because of the absent air damping. The process of measurement was extremely simple without the 3D translation stage, what we need to do was just putting the glass plate into the vacuum cavity carefully.

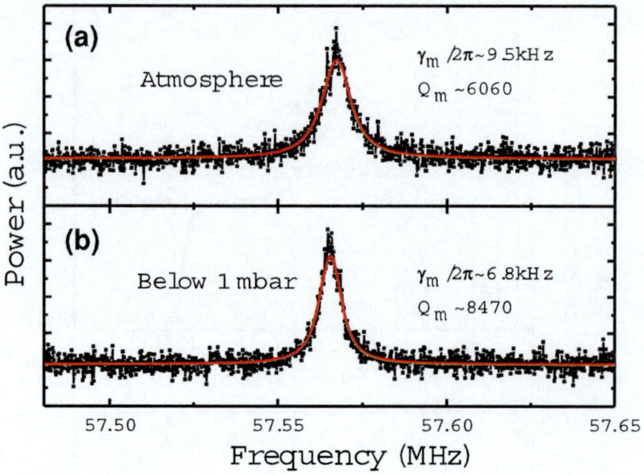


**Fig. 6.5** **a–b** are power spectrum of the mechanical vibrations in atmospheric pressure and in vacuum for the packaged microresonator system, respectively

## 6.4 Summary

In conclusion, we propose a method to package the silica microsphere-fiber coupling system for optomechanical study. We pay close attention to keep high mechanical $Q$ factor and simultaneously keep ultrahigh optical $Q$ factor. In the experiment, we discover that the packaged coupling system which is operated on 780 nm wavelength can achieve this goal. Because 780 nm wavelength can match up with the waist size of the tapered fiber which is small enough to reduce the mechanical clamping losses.

The experiment in vacuum is also carried out to verify the practicability of the packaged system. The result shows that the touched mechanical $Q$ factor can be improved without the air damping. This work will not only make the experiments on microcavity optomechanics convenient and low cost in vacuum and cryostat, but pave the way to the sensing and quantum information applications of cavity optomechanical oscillation.

## References

1. Zou CL, Shu FJ, Sun FW, Gong ZJ, Han ZF, Guo GC (2013) Opt Express 21(8):9982
2. MacDonald A, Popowich G, Hauer B, Kim P, Fredrick A, Rojas X, Doolin P, Davis J (2015) Rev Sci Instrum 86(1):013107
3. Yan YZ, Zou CL, Yan SB, Sun FW, Ji Z, Liu J, Zhang YG, Wang L, Xue CY, Zhang WD et al (2011) Opt Express 19(7):5753
4. Wang P, Ding M, Lee T, Murugan GS, Bo L, Semenova Y, Wu Q, Hewak D, Brambilla G, Farrell G (2013) Appl Phys Lett 102(13):131110
5. Wang P, Ding M, Murugan GS, Bo L, Guan C, Semenova Y, Wu Q, Farrell G, Brambilla G (2014) Opt Lett 39(17):5208

6. Vanier F, Peter YA, Rochette M (2014) Opt Express 22(23):28731
7. Dong Y, Wang K, Jin X (2015) Appl Opt 54(2):277
8. Monifi F, Ozdemir SK, Friedlein J, Yang L (2013) IEEE Photonics Technol Lett 25(15):1458
9. Park YS (2009) Radiation pressure cooling of a silica optomechanical resonator. PhD thesis
10. Chen Y, Shen Z, Xiong X, Dong CH, Zou CL, Guo GC (2016) New J Phys 18(6):063031
11. Carmon T, Yang L, Vahala KJ (2004) Opt Express 12(20):4742
12. Kippenberg T, Spillane S, Vahala K (2004) Phys Rev Lett 93(8):083904
13. Kippenberg TJ, Rokhsari H, Carmon T, Scherer A, Vahala KJ (2005) Phys Rev Lett 95(3):033901
14. Riviere R, Arcizet O, Schliesser A, Kippenberg TJ (2013) Rev Sci Instrum 84(4):043108
15. Anetsberger G, Riviere R, Schliesser A, Arcizet O, Kippenberg TJ (2008) Nat Photonics 2(10):627
16. Kippenberg TJ, Vahala KJ (2008) Science 321(5893):1172
17. Knight JC, Cheung G, Jacques F, Birks T (1997) Opt Lett 22(15):1129
18. Zou CL, Yang Y, Dong CH, Xiao YF, Wu XW, Han ZF, Guo GC (2008) JOSA B 25(11):1895
19. Little BE, Laine JP, Haus HA (1999) J Lightwave Technol 17(4):704
20. Dong C, Zou C, Cui J, Yang Y, Han Z, Guo G (2009) Chin Opt Lett 7(4):299

# Chapter 7
# Phase Sensitive Imaging of Mechanical Modes

**Abstract** The characterization of vibrational modes of electro/optomechanical devices, including identification of the vibration frequencies, profiles, mode losses and so on, is important for validating their design, operation and performance. Scanning optical vibrometer enable high-resolution, non-invasive mapping of the vibration pattern of acoustic devices. It also offers complete quadrature information rather than pure intensity response. We demonstrate a high frequency phase-sensitive heterodyne vibrometer, operating up to 10 GHz. Using this heterodyne vibrometer, the amplitude and phase fields of the fundamental thickness mode, the radial fundamental and the 2nd-order modes of an AlN optomechanical microdisk resonator are mapped with a displacement sensitivity of around $0.36\ \mathrm{pm}/\sqrt{\mathrm{Hz}}$. The simultaneous amplitude and phase measurement allows precise mode identification and characterization. The recorded modal frequencies and profiles are consistent with numerical simulations. This vibrometer will be of great significance for the development of high frequency mechanical devices.

## 7.1 Introduction

In recent years, the continued progresses of micro/nano optomechanical systems and micro/nano electromechanical systems (MEMS/NEMS) highlight the importance of vibration freedom on the microscale. The quantization of the elastic wave or vibration mode in solid is called phonon. By measuring sound waves and vibrations in various materials and structures, the physical properties of materials (sound velocity, wavelength, etc.) and the binding or scattering characteristics of sound waves by micro/nano structures can be obtained, which promotes the development of elastic wave dynamics. At the same time, phonons can travel a long distance in solids, thus they can also be used as information carriers and highly sensitive sensors. Because mechanical systems are easily coupled with various physical systems, they have also been deeply studied in the quantum field [1, 2], including ground-state cooling of mechanical resonators [3–5], state transfer between phonons and spins of quantum dots or diamond color centers [6], strong coupling between mechanical oscillator and microwave/optical resonators [7–9], efficient conversion between photons at vastly



Z. Shen, *Experimental Research of Cavity Optomechanics*, Springer Theses,
https://doi.org/10.1007/978-981-33-4458-7_7

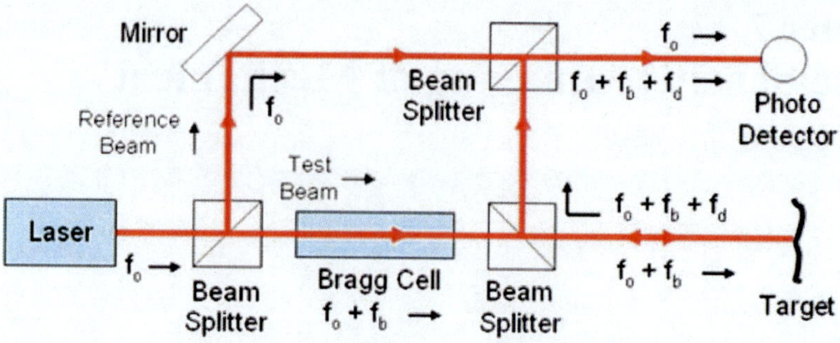


**Fig. 7.1** Schematic of the laser Doppler vibrometer

different frequencies [10, 11], and non-reciprocal optomechanical devices [12–14]. Particularly, high frequency mechanical devices above 10 GHz are desirable for advancing mechanical quantum devices because they require less stringent refrigeration conditions for reaching quantum mechanical ground state and their frequency are matched to superconducting quantum circuits. A 10 GHz mechanical oscillator can be cooled to ground state by direct dilution refrigeration without the necessity of active cooling techniques. Furthermore, the extension of acoustic resonators to microwave X-band and beyond is important for next generation wireless communications [15–17].

A commonly used instrument non-contactly measuring the surface vibration of an object is the laser vibrometer, which can detect the motion of objects that are difficult to access or have a small size and high temperature. Vibrometer are widely used in aerospace, automotive design, data storage of hard disk. The principle of the current typical laser vibrometer relies on the Doppler effect, as shown in Fig. 7.1. When the laser is irradiated perpendicularly to the surface of the vibrating object, the reflected light frequency will shift due to the Doppler effect, and the frequency shift is proportional to the surface vibration rate. The reflected light and the incident laser are sent to the detector to obtain the beat signal of the probe beam and the reference beam, and then the vibration signal of the object surface is obtained by demodulating the beat signal. For the macro-scale object, the vibration frequency is relatively low and the amplitude is large. Therefore, the requirements of the performances of laser vibrometer are relatively low. The existing commercial products can basically meet the measurement requirements (Fig. 7.2).

On the other hand, the vibration modes in the micro/nano mechanical structure have the characteristics of high frequency, small wavelength, and low quantum thermal noise, which enable the acoustic components easier to integrate and obtain higher signal-to-noise ratio. Especially in the quantum regime, its importance is becoming increasingly prominent. However, detecting high-frequency vibrations in micro/nano systems requires laser vibrometers to have larger measurement bandwidth, higher displacement and spatial resolution. In addition, optomechanical systems require low temperature environment for quantum applications. Therefore, how to monitor

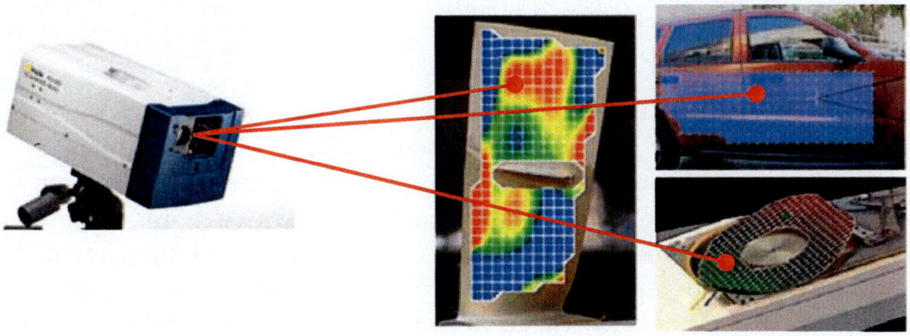

**Fig. 7.2** The applications of laser vibrometer

the operating status of devices in low temperature environments or other complex environments also needs to be solved.

In 2000, Knuuttila et al. demonstrated two-dimensional imaging of 950 MHz surface acoustic wave using a scanning homodyne Michelson interferometer [18]. In 2001, Graebner et al. described a stabilized homodyne interferometer for accurate imaging of 1.8 GHz surface vibrations of a thin-film resonator [19]. In 2007, Martinussen et al. and Kokkonen and Kaivola demonstrated a heterodyne interferometer, which facilitated measurement of absolute amplitude and phase of the surface vibration without calibration [20, 21]. They also performed measurements of micromechanical devices in the frequency range 10–950 MHz. In 2013, Erlend et al. described modifications to the heterodyne interferometer and improved the displacement sensitivity to 7.1 $\text{pm}/\sqrt{\text{Hz}}$ at 21 MHz [22]. In 2014, Hashimoto et al. demonstrated the detection of surface acoustic wave and bulk acoustic wave devices with 5 GHz vibration frequency [23].

Although great sensitivities have been achieved, the frequencies of the vibration modes mapped by above methods are limited to below a few gigahertz. Optical pump-probe technique, which requires high-performance pulsed laser, allows the time-domain response of mechanical oscillators to be measured. Frequency-domain measurements with continuous laser, as a complementary method, is often used to characterize mechanical resonators with high quality factors. Here, we demonstrate a scanning heterodyne vibrometer for realizing simultaneous absolute amplitude and phase imaging of vibration up to 10 GHz-range in an AlN microdisk resonator with a displacement sensitivity of 0.36 $\text{pm}/\sqrt{\text{Hz}}$ for out-of-plane motion detection. Measurements on the fundamental thickness mode, the fundamental radial mode and the 2nd-order radial modes of an AlN microdisk provide phase-sensitive modal visualization that is consistent with our numerical simulation.

## 7.2 Experimental Setup

The vibrometer derives the vibration from the modulated light signal by a heterodyne interferometer. The experimental setup is shown in Fig. 7.3. A linearly polarized 633 nm HeNe laser is split into two paths—the reference beam and the probe beam—by a polarizing beam splitter (PBS). A half-wave plate ($\lambda/2$) is used to tune the power ratio between the reference and the probe beams. The frequency of the reference beam is shifted from the original input laser frequency $\omega$ to $\omega + \omega_{\text{ref}}$, by an acousto-optic modulator (AOM). The probe beam is focused perpendicularly onto the sample with a microscope objective (NA = 0.55) and reflected back to the PBS. The polarization of the reflected probe signal is rotated by 90° because of the double pass through the $\lambda/4$ plate, which leads to the same polarization as the frequency-shifted reference light. Then the motion-modulated probe and reference light are combined with the beam splitter and sent into a photodetector (PD) through a single mode fiber. The fiber acts as a confocal pinhole to increase the lateral resolution. The theoretical resolution limit in our setup is around 600 nm ($0.5\lambda/\text{NA}$) [24].

To analyze the heterodyne signal, we consider the sample surface vibrating sinusoidally as $Z_0 \sin(\Omega t + \phi)$ in the z-direction (perpendicular to the surface of sample),

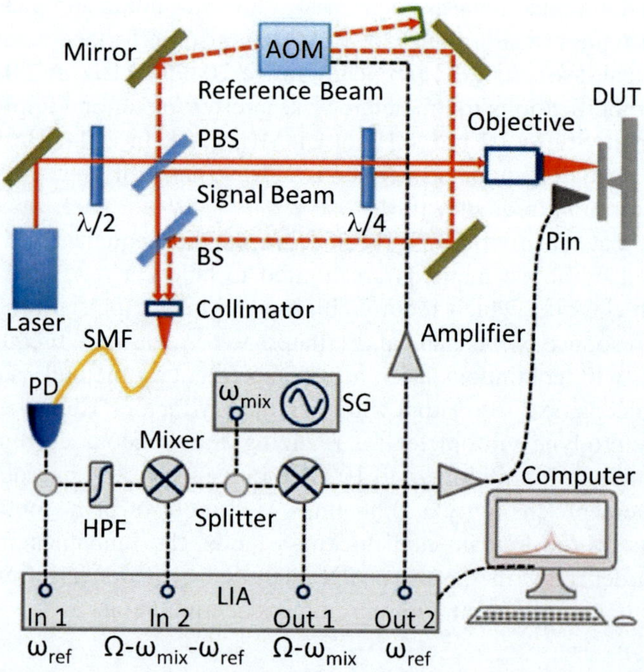


**Fig. 7.3** Schematic layout of the heterodyne vibrometer. AOM: acousto-optic modulator, PBS: polarizing beam splitter, BS: beam splitter, DUT: device under test, Pin: dipolar pin antenna actuator, SMF: single mode fiber, PD: photodetector, HPF: high-pass filter, SG: signal generator, LIA: lock-in amplifier

where $Z_0$, $\Omega$ and $\phi$ are the amplitude, angular frequency and phase of the vibration, respectively. The probe light field detected by the PD is $E_p = A_1\cos[\omega t + \phi_1 + 2kZ_0 sin(\Omega t + \phi)]$ and the reference light field is $E_{\text{ref}} = A_2\cos[(\omega + \omega_{\text{ref}})t + \phi_2]$, where $A_1$ ($A_2$) and $\phi_1$ ($\phi_2$) denote the amplitudes and phases of the probe (reference) field, and $k = 2\pi/\lambda$ is the optical wavenumber. When $Z_0 \ll \lambda$, the $E_{\text{p}}$ can be expanded to the first order of $Z_0$ as

$$\begin{aligned} E_{\text{p}} \approx\ & A_1\cos(\omega t + \phi_1) + A_1 k Z_0 \cos[(\omega + \Omega)t + \phi_1 + \phi)] \\ & - A_1 k Z_0 \cos[(\omega - \Omega)t + \phi_1 - \phi)] \end{aligned} \tag{7.1}$$

The PD response to the intensity of light field is

$$\begin{aligned} I(t) &= \alpha \left| E_{\text{p}} + E_{\text{ref}} \right|^2 \\ &\approx \alpha\{A_1 A_2 \cos(\omega_{\text{ref}} t + \phi_2 - \phi_1) \\ &\quad + A_1 A_2 k Z_0 \cos[(\Omega - \omega_{\text{ref}})t + \phi + \phi_1 - \phi_2]\} \end{aligned} \tag{7.2}$$

where $\alpha$ is the responsivity of the PD. Here, we have dropped the second-order term of $Z_0^2$ and also filtered out the component oscillating at high frequency $\Omega + \omega_{\text{ref}}$. When the lock-in amplifier (LIA) demodulates at $\omega_{\text{ref}}/2\pi$ and $\omega_{\text{sig}}/2\pi = (\Omega - \omega_{\text{ref}})/2\pi$ simultaneously, the amplitude $R_{\text{ref}} = A_1 A_2$ ($R_{\text{sig}} = k Z_0 R_{\text{ref}}$) and phase $\theta_{\text{ref}} = \phi_2 - \phi_1$ ($\theta_{\text{sig}} = \phi + \phi_1 - \phi_2$) of the reference (signal) component are determined. Then, the amplitude and the phase of the surface vibration can be calculated as:

$$\begin{aligned} Z_0 &= \frac{1}{k}\frac{R_{\text{sig}}}{R_{\text{ref}}} \\ \phi &= \theta_{\text{sig}} + \theta_{\text{ref}} \end{aligned} \tag{7.3}$$

External disturbances, such as temperature fluctuations and instabilities of laser wavelength and power, may lead to individual drift of the amplitude $A_1$ ($A_2$) and phase $\phi_1$ ($\phi_2$) of the probe (reference) light field, which give rise to drifts of signal component $R_{\text{sig}}$ and $\theta_{\text{sig}}$. However, this can be compensated by measuring the reference component $R_{\text{ref}}$ and $\theta_{\text{ref}}$. Thus, the heterodyne vibrometer is insensitive to such external disturbances.

In our experiment, the frequency of signal $\omega_{\text{sig}}/2\pi$ should be mixed with a signal generator (SG) to a lower frequency that falls within the frequency operation range of the lock-in amplifier LIA (Zurich instruments UHFLI), i.e., less than 600 MHz. Therefore, the detector signal from the PD (Newport, 1580-A, 12 GHz) is split by a power splitter with one of the outputs sent through a high-pass filter (HPF) to filter out the reference component, then mixed down to the new frequency $\omega'_{\text{sig}} = \Omega - \omega_{\text{ref}} - \omega_{\text{mix}}$, $\omega_{\text{mix}}$, where $\omega_{\text{mix}}$ denotes the operating frequency of the SG. The mixer output is fed into the LIA and demodulated at $\omega'_{\text{sig}}/2\pi$. The amplitude of demodulation signal is $R'_{\text{sig}} = \xi R_{\text{sig}}$, where $\xi$ is a conversion factor determined by the conversion loss of mixer, insertion loss of HPF and coaxial cables, and imbalance of the power splitter. It should be pointed out that this conversion factor is nonconstant when we measure different mechanical modes in the AlN microdisk due to the frequency-dependent response of electronic components. For the measurement at around 10.4 GHz, 1.67 GHz and 635 MHz as will be described below, $\xi$ is around 1/4.87, 1/2.85, and 1, respectively. The signal output 1 of LIA is set at $\Omega - \omega_{\mathrm{mix}}$ and mixed with the SG to actuate the device using a dipolar pin antenna. The signal output 2 of LIA is used to drive the AOM. The SG is clock-referenced to the LIA to ensure that they are phase locked. The device under test is placed on a linear piezo stage (Micronix PPS-20) with lateral resolution of 10 nm, which allows two-dimensional point-by-point scanning.

Here, we focus on the noise analysis for the measurement frequency around 10.4 GHz. The displacement sensitivity is defined as the normalized minimum detectable vibration amplitude $A_{\mathrm{min,total}} = \frac{1}{k}\frac{S_{\mathrm{total}}^{1/2}}{\xi R_{\mathrm{ref}}}$, where $R_{\mathrm{ref}}$ (around 13 mV) is measured by the LIA and $S_{\mathrm{total}}^{1/2}$ is the total noise. $S_{\mathrm{total}}^{1/2}$ includes contribution from all available noise sources: electrical noise (noise without light input) $S_{\mathrm{PD}}^{1/2}$ from the PD, shot noise $S_{\mathrm{shot}}^{1/2}$ due to the discrete nature of photon, and the noise from other electrical parts (including LIA, amplifier and so on). With the PD connected, and with no incident light, the PD noise is measured as $S_{\mathrm{PD}}^{1/2} = 9\,\mathrm{nV}/\sqrt{\mathrm{Hz}}$, which corresponds to a measurement sensitivity $A_{\mathrm{min,PD}} = \frac{1}{k}\frac{S_{\mathrm{PD}}^{1/2}}{\xi R_{\mathrm{ref}}} = 0.34\,\mathrm{pm}/\sqrt{\mathrm{Hz}}$. The shot noise is given by $S_{\mathrm{shot}}^{1/2} = \frac{1}{2} \times \sqrt{2e\alpha P_{\mathrm{in}} B} \times G$, where $e$ is the elementary charge, $P_{\mathrm{in}}$ the incident laser power on the PD, $B$ the detection bandwidth, and $G$ the gain of build-in transimpedance amplifier of the PD. In our measurements, $\alpha = 0.22$ A/W, $P_{\mathrm{in}} = 200\,\mu$W, $B = 1$ Hz, and $G = 3$ V/mA. The prefactor $\frac{1}{2}$ is attributed to the voltage division between the 50 Ω output impedance of the PD and the 50 Ω input impedance of the LIA [25]. Therefore, the resulting shot-noise limited sensitivity becomes $A_{\mathrm{min,shot}} = \frac{1}{k}\frac{S_{\mathrm{shot}}^{1/2}}{\xi R_{\mathrm{ref}}} = 0.015\,\mathrm{pm}/\sqrt{\mathrm{Hz}}$. The sum of electrical noise from the PD and the shot noise ($\sqrt{A_{\mathrm{min,PD}}^2 + A_{\mathrm{min,shot}}^2}$) gives a theoretical sensitivity limit of $0.34\,\mathrm{pm}/\sqrt{\mathrm{Hz}}$, which is close to our measured value $0.36\,\mathrm{pm}/\sqrt{\mathrm{Hz}}$. The above analysis shows that the PD noise is the dominant noise source. For the measurement frequency of 1.67 GHz and 635 MHz, the sensitivity is around $0.33\,\mathrm{pm}/\sqrt{\mathrm{Hz}}$, which is consistent with the above conclusion.

## 7.3 Imaging Results and Analysis

The aluminum nitride microdisk cavity used in this article is usually prepared by two-step photolithography and dry etching [26–29], as shown in Fig. 7.6. In the first step of electron beam lithography, the mask uses a hydrogen silsesquioxane (HSQ) photoresist to determine the optical structure pattern, including the micro-disk cavity, the coupled waveguide, and the coupled grating; then the chlorine-based dry etching is used to etch the AlN layer to 70 nm thick; in the second step of electron beam lithography, the mask uses ZEP520A photoresist, and the circular window for the next etching of the oxide layer is determined at the edge of the microdisk cavity; at this time, in the first step of lithography, the remaining HSQ still serves as a mask to protect the microdisk cavity and waveguide, ensuring that only the AlN layer of the circular window is etched, and the remaining 70 nm thick AlN outside the circular window will be retained and used as a mask for the next wet etching. The last step is to determine the wet etching time to ensure that the microdisk cavity is suspended and the support column is thin enough.

Figure 7.4a shows the SEM image of the device. A suspended AlN microdisk with radius 5.5 μm and thickness 550 nm is fabricated adjacent to an optical waveguide. The radius of the supporting pedestal is controlled to be around 100 nm. The photonic structure is patterned using the two-step e-beam lithography and dry etching technique. We actuate the microdisk piezoelectrically using a dipolar pin antenna and read out the mechanical motion optically through measuring the optical transmission signal of the optical whispering-gallery mode, as shown in Fig. 7.4a. Figure 7.4b shows a broad-band driven response spectrum. The radial modes can be distinctly identified up to the 5th-order at 4.5 GHz and the fundamental thickness mode at

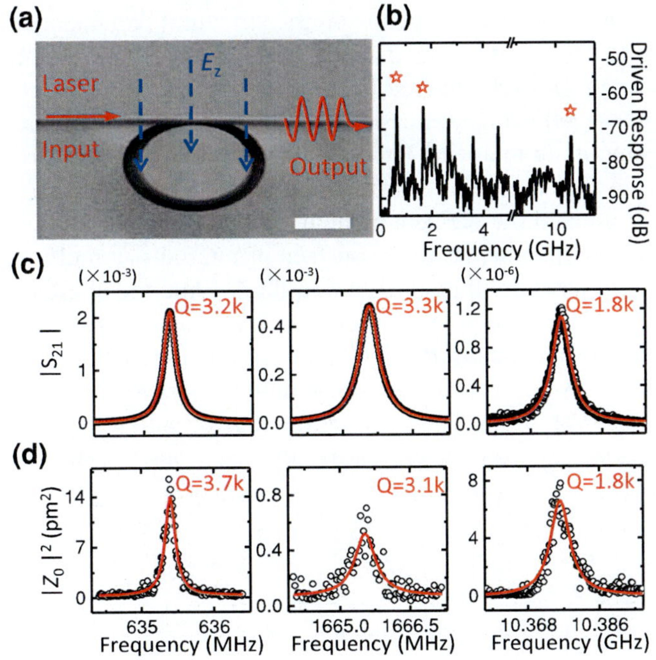


**Fig. 7.4** **a** SEM image of a microdisk resonator of 5.5 μm in radius coupled with optical waveguide. The scale bar is 5 μm. **b** Piezo-electric driven response of the disk transduced by the integrated waveguide. The red stars mark the resonance peaks of the fundamental and the second-order radial modes and the fundamental thickness mode. **c** Zoom-in spectra of the three modes at 635.4 MHz, 1665.5 MHz, and 10.37 GHz. **d** The corresponding resonances measured using the heterodyne vibrometer

10.37 GHz. Zoomed-in spectra of three modes are shown in Fig. 7.4c. The quality factors measured are fitted to be around 3200, 3300 and 1800, respectively.

Figure 7.4d shows the frequency responses obtained by our heterodyne interferometer where the probe laser spot is focused on the center of the microdisk and the signal generator operating frequency $\omega_{\mathrm{mix}}$ is scanned with a fixed electric excitation of around 25 dBm for the fundamental and 2nd-order radial modes and 20 dBm for fundamental thickness mode. As shown in Fig. 7.4d, the vibrometer measurements generally agree with optomechanically-transduced responses in Fig. 7.4c. Due to reduced out-of-plane displacement, the third-order and above radial modes are not detected using our heterodyne interferometer.

The demodulated two-dimensional (2D, $16 \times 16\,\mu\mathrm{m}^2$) vibration amplitude ($Z_0$) and phase ($\phi$) data of the fundamental thickness mode (10.37 GHz) are presented in Fig. 7.5a, f. The strong signal at the edge of the microdisk is due to the multi-path interference of light bouncing back at different sample heights. As illustrated in Fig. 7.5b, when the laser spot is fixed at this position, the probe light field became $E'_p = A_1\cos[\omega t + \phi_1 + 2kZ_0\cos(\Omega t + \phi)] + A_3\cos[\omega t + \phi_3]$, where the $A_3$ and $\phi_3$ denote the amplitude and phase of the reflected probe field by the silicon substrate. The demodulated signal $R_{\mathrm{ref}}$ of LIA is determined by the relative phase $\phi_3 - \phi_1$, as well as $A_1$ and $A_3$. The phase depends on the structural parameters (2.2-μm-thick $SiO_2$ buffer layer, 550-nm-thick microdisk) and $A_1$ ($A_3$) can be directly measured in the experiment. As shown by the red line in Fig. 7.5c, the fitting by considering this edge effect agrees well with the experiment results (black dots). Figure 7.5e plots the error-filtered amplitude image of Fig. 7.5a. Similar analyses are performed for the other vibration modes. Figure 7.5d displays the simulated thickness mode shape using the three-dimensional finite-element method (COMSOL Multiphysics). For the out-of-plane motion (perpendicular to the surface of microdisk), multi-turn concentric rings are found in the amplitude distribution field, while the phase field is uniform except the center where the pillar supports the microdisk, as shown in Fig. 7.5g, h, respectively. Because of the limited lateral resolution (around 900 nm for our setup), the spatial images we detect are the convolution transformation of the vibration fields, as shown in Fig. 7.5i, j. The convolution transformation formula is $Z'(x, y) = 0.57 \iint dx'dy' \cdot Z_{sim}(x', y') \cdot exp(-\frac{(x'-x)^2+(y'-y)^2}{r^2/2})$, where $r = 900$ nm is the radius of the Gaussian-shape spot, $Z'$ is the expected displacement of measurement, $Z_{sim}$ is the simulated displacement, and 0.57 is the normalization factor. The simulated convolution images of amplitude and phase response match well the experiment results (Fig. 7.6).

Figures 7.7a and 7.8a show the simulation results of first-order and second-order radial vibration modes, whose resonant frequencies are 635 MHz and 1.6 GHz, respectively. The measured amplitude and phase fields are plotted in Figs. 7.7b and 7.8b with the simulated of out-of-plane motion shown in Figs. 7.7c and 7.8c. Benefiting from the 2D images of acoustic field distribution, especially the phase information, we can identify the observed modes unambiguously. For both vibrational modes, the energy of out-of-plane motion concentrates in the central part of the microdisk, we can speculate that the elastic wave leakage at the pedestal will induce

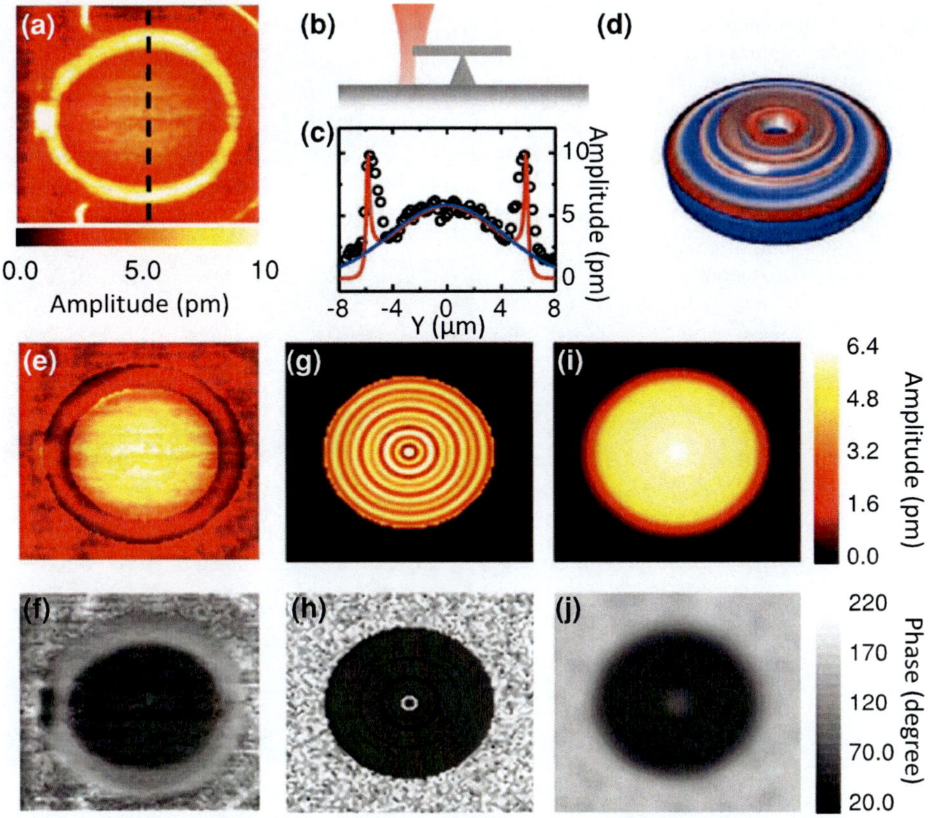


**Fig. 7.5** **a** Measured amplitude distribution of the thickness mode excited at a 10.37 GHz drive with a detection bandwidth of 10 Hz. **b** The laser spot locates at the edge of the microdisk. Due to the light diffraction, both the AlN microdisk and silicon substrate contribute to the reflected light. **c** Black dots indicate the measured amplitude distribution along the dash black line in **a**. The blue line and red line denote the calculated amplitude distribution before and after the introduction of the measurement error at the edge of the microdisk. **d** The simulated pattern of the thickness mode. **e** Error-filtered amplitude image of **a**. **f** Measured phase distribution of the thickness mode. **g–h** The simulated amplitude and phase fields of the thickness mode, respectively. **i–j** The simulated images after convolution with the laser spot radius of 900 nm

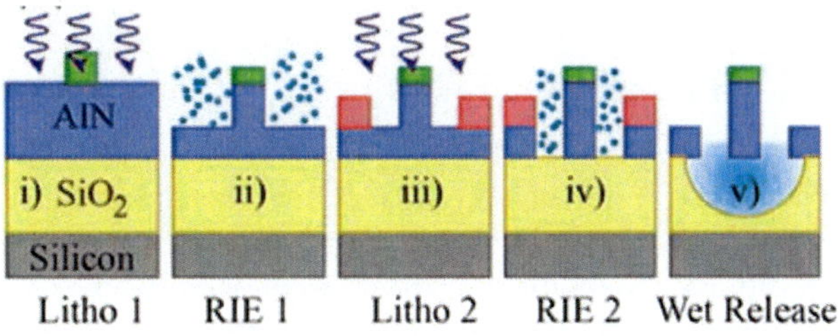


**Fig. 7.6** Schematic diagram of AlN microdisk cavity preparation process. Reprinted with permission from Ref. [26]. Copyright 2003 by American Institute of Physics

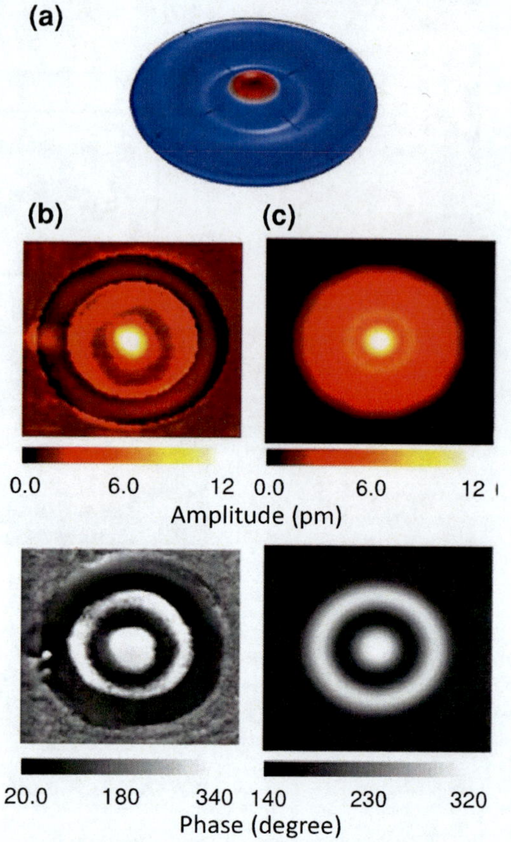


**Fig. 7.7** **a** Simulated displacement profiles of the fundamental radial mode. **b** The measured amplitude and phase fields of the fundamental radial mode. **c** The simulated amplitude and phase fields of the out-of-plane motion for the fundamental radial mode

significant mechanical loss for radial modes. Therefore, reducing the size of the supporting pedestal can improve the $Q$ factor of the mechanical modes. Simulation results show that the $Q$ factors are proportional to $\sigma^{-1.1}$ for the fundamental radial mode and $\sigma^{-2.1}$ for the second-order radial mode, respectively, where $\sigma$ is the surface area of the supporting pedestal.

## 7.4 Summary

We have developed a high frequency scanning heterodyne vibrometer, which can measure the mechanical motion up to 10 GHz. It not only produces high-resolution amplitude distribution of the vibration modes but also clearly maps out the phase of the vibration. This unique capability allows us to fully characterize the thickness mode and radial modes supported by the AlN microdisk and make accurate comparisons with numerical simulations. The phase-sensitive heterodyne vibrometer demonstrated in our paper is a powerful tool for characterizing mechanical modes

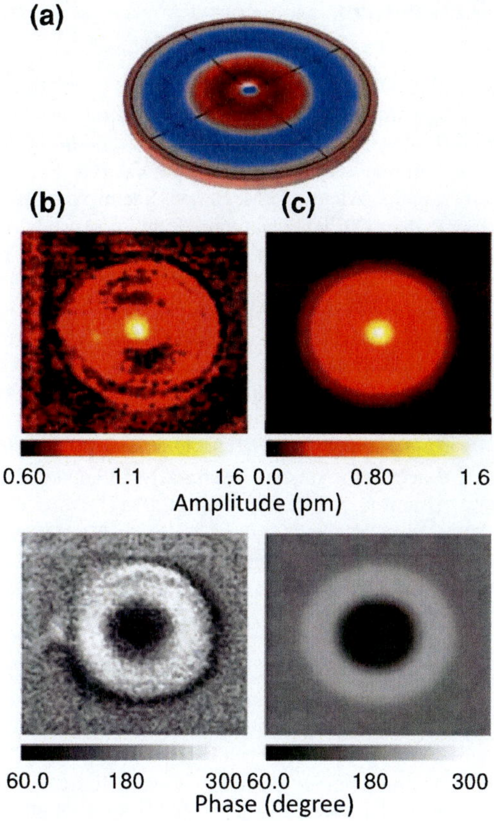


**Fig. 7.8** **a** Simulated displacement profiles of the second-order radial mode. **b** The measured amplitude and phase fields of the second-order radial mode. **c** The simulated amplitude and phase fields of the out-of-plane motion for the second-order radial mode

in micromechanical components, and enables better design and performance of high frequency optomechanical and electromechanical devices.

In conclusion, our laser vibrometer has the following advantages: 1. The vibrometer is based on heterodyne interferometry. The intensity and phase information of the vibration is obtained by comparing the reference field and the detection signal. Not only can the absolute amplitude be directly obtained, but can also avoid many unfavorable disturbances, such as temperature drift in the arms of the interferometer, so as to obtain accurate and reliable intensity and phase information; 2. The photoelectric signal of the vibrometer is demodulated by a lock-in amplifier, where the amplitude and phase information can be directly obtained through simple calculations. Thus the weak signal can be measured to achieve high sensitivity; 3. For measuring mechanical vibrations of different frequencies, the vibrometer can be reconfigured to use of different bandwidth optical detectors. For low-frequency detection, a low-bandwidth photodetector can be used, whose lower noise can improve the detection resolution of the vibrometer; 4. The detection optical path of the vibrometer can use the principles of confocal microscopy, where the spot of the detection laser beam can be focused to approach the theoretical limit, thereby improving the lateral spatial resolution of the vibrometer.

## References

1. Regal C, Lehnert K (2011) J Phys: Conf Ser (IOP Publishing, 2011) 264:012025
2. Aspelmeyer M, Kippenberg TJ, Marquardt F (2014) Rev Mod Phys 86(4):1391
3. Teufel J, Donner T, Li D, Harlow J, Allman M, Cicak K, Sirois A, Whittaker JD, Lehnert K, Simmonds RW (2011) Nature 471:204
4. Chan J, Alegre TPM, Safavi-Naeini AH, Hill JT, Krause A, Groeblacher S, Aspelmeyer M, Painter O (2011) Nature 478:89
5. Peterson R, Purdy T, Kampel N, Andrews R, Yu PL, Lehnert K, Regal C (2016) Phys Rev Lett 116(6):063601
6. Golter DA, Oo T, Amezcua M, Stewart KA, Wang H (2016) Phys Rev Lett 116(14):143602
7. Teufel JD, Li D, Allman MS, Cicak K, Sirois AJ, Whittaker JD, Simmonds RW (2011) Nature (London) 471:204
8. Verhagen E, Deleglise S, Weis S, Schliesser A, Kippenberg TJ (2012) Nature 482:63
9. Han X, Zou CL, Tang HX (2016) Phys Rev Lett 117(12):123603
10. Dong C, Fiore V, Kuzyk MC, Wang H (2012) Science 338(6114):1609
11. Bagci T, Simonsen A, Schmid S, Villanueva LG, Zeuthen E, Appel J, Taylor JM, Sorensen AS, Usami K, Schliesser A et al (2014) Nature 507:81
12. Dong CH, Shen Z, Zou CL, Zhang YL, Fu W, Guo GC (2015) Nat Commun 6:6193
13. Fu W, Shu FJ, Zhang YL, Dong CH, Zou CL, Guo GC (2015) Opt Express 23(19):25118
14. Shen Z, Zhang YL, Chen Y, Zou CL, Xiao YF, Zou XB, Sun FW, Guo GC, Dong CH (2016) Nat Photonics 10:657
15. Satoh Y, Nishihara T, Yokoyama T, Ueda M, Miyashita T (2005) Jpn J Appl Phys 44(5R):2883
16. Tadesse SA, Li M (2014) Nat Commun 5:5402
17. Li H, Tadesse SA, Liu Q, Li M (2015) Optica 2(9):826
18. Knuuttila JV, Tikka PT, Salomaa MM (2000) Opt Lett 25(9):613
19. Graebner J, Barber B, Gammel P, Greywall D, Gopani S (2001) Appl Phys Lett 78(2):159
20. Martinussen H, Aksnes A, Engan HE (2007) Opt Express 15(18):11370
21. Kokkonen K, Kaivola M (2008) Applied Physics Letters 92(6):063502
22. Leirset E, Engan HE, Aksnes A (2013) Optics express 21(17):19900
23. K.y. Hashimoto, S. Kawachi, A. Takahashi, S. Sakamoto, T. Omori, in *Frequency Control Symposium (FCS), 2014 IEEE International* (IEEE, 2014), pp. 1–4
24. Rembe C, Drabenstedt A (2006) Review of scientific instruments 77(8):083702
25. Whitman R, Korpel A (1969) Applied Optics 8(8):1567
26. Pernice WHP, Xiong C, Schuck C, Tang HX (2012) Applied Physics Letters 100(9):091105
27. Xiong C, Sun X, Fong KY, Tang HX (2012) Applied Physics Letters 100(17):171111
28. Han X, Xiong C, Fong KY, Zhang XF, Tang HX (2014) New Journal of Physics 16(6):063060
29. Han X, Fong KY, Tang HT (2015) Applied Physics Letters 106(16):161108